Demography and Evolution
in Plant Populations

BOTANICAL MONOGRAPHS

BOTANICAL MONOGRAPHS · VOLUME 15

Demography and Evolution in Plant Populations

EDITED BY

OTTO T. SOLBRIG

Department of Biology and
Gray Herbarium
Harvard University

UNIVERSITY OF CALIFORNIA PRESS

BERKELEY AND LOS ANGELES 1980

UNIVERSITY OF
CALIFORNIA PRESS

ISBN: 0 520 03931 9

Library of Congress
Catalog Card Number: 79-64486

© 1980 Blackwell Scientific Publications

Printed in Great Britain by
The Alden Press, Oxford
and bound by
Kemp Hall Bindery, Oxford

To Dorothy

Contents

Contributors

WARREN G. ABRAHAMSON *Department of Biology, Bucknell University, Lewiston, Pennsylvania, USA*

ROBERT COOK *Department of Biology, Harvard University, Cambridge, Massachusetts, USA*

DAVID G. LLOYD *Department of Botany, University of Canterbury, Christchurch 1, New Zealand*

JOSÉ SARUKHÁN *Departmento de Botanica, Instituto de Biologia, Universidad Nacional Autónoma de Mexico, Mexico 20, Distrito Federal, Mexico*

ROY W. SNAYDON *Agricultural Botany Department, The University, Reading, United Kingdom*

OTTO T. SOLBRIG *Department of Biology and Gray Herbarium, Harvard University, Cambridge, Massachusetts USA*

JAMES WHITE *Department of Botany, University College, Dublin 4, Ireland*

Foreword

HERBERT G. BAKER

In the last decade, there has been a growing realization by botanists that demography need not, indeed should not, be a subject that deals only with the dynamics of human and other animal populations. However, John L. Harper and James White, in writing a review article in 1974, had considerable difficulty in finding in the literature quantitative data about the growth, maintenance (and decline) of plant populations. Nevertheless, their review (*Annual Review of Ecology and Systematics* 5:419–463) 'The Demography of Plants' undoubtedly started many investigators working to supply such data, as well as setting them thinking about principles and searching the literature for hidden gems. Now, the emergent discipline of plant demography has reached the book-production stage.

Two kinds of approach may be made to a book on any new subject. A particularly well-informed investigator (who is also a talented writer) may undertake the monumental task of writing a textbook on his own. John Harper has done this successfully, recently. His book *Population Biology of Plants* (London, Academic Press, 1977) is a magnificent achievement. But it must also be recognized that consideration of the subject from the different points of view of a group of authors can add significantly to our appreciation of it. This is the rationale of the present work and its inclusion in the 'Botanical Monographs' series.

Otto T. Solbrig has brought together an outstanding team of seven contributors with experience in more than that number of areas of the subject and in many parts of the world. Each of these botanists has made substantial firsthand contributions to the subject and, as a group, they are particularly well equipped to survey the field, draw conclusions and suggest paths that plant demographic investigations might follow in the future.

A strength of this book is the evolutionary awareness of its authors. They are not content to view demography as a means of interpreting the *status quo*. The evolutionary context in which all demography must be placed eventually is brought out by Solbrig in Chapter 1 and is kept in mind throughout the book. The logical order in which the chapters are arranged promotes a flow of thought in this vein. Another important innovation is the inclusion of José

Sarukhán's chapter on demographic problems in tropical systems. Roy W. Snaydon covers the demography of agricultural systems which are no less important for being more or less under human control.

James White considers the impact of demographic factors in populations of plant, while Otto T. Solbrig demonstrates the relevance of various gene systems to the genetic structure of plant populations. David G. Lloyd draws on many years of concern with the relationship between breeding systems of flowering plants and their population structures, while Warren G. Abrahamson considers the effects that vegetative reproduction may have on the picture. Flowering plant populations differ from most animal populations in that, at any time, a greater or lesser proportion of the population may be in a dormant (seed-reservoir) condition; Robert Cook considers this and other aspects of the biology of seeds in soil.

Thus, this book is wide ranging in its consideration of flowering plant demography, and it takes its place appropriately in the 'Botanical Monographs' series. It is to be hoped that future volumes in this series will cover the demography of many other kinds of plants besides the flowering plants that are so well treated here.

Botany Department, University of California,
Berkeley, California 94720

Chapter 1
Demography and Natural Selection

OTTO T. SOLBRIG

Demography deals with the quantitative aspects of birth, growth, reproduction and death in a population. Originally the term demography was applied exclusively to the study of human populations. By analogy it was applied later in the ecological literature to the study of other species of animals and more recently also to the study of plants (Harper & White 1974).

Collecting demographic data on human populations is economically and politically important. Governments regularly gather a variety of such statistics and maintain special bureaucracies with highly trained personnel and expensive and advanced machinery for the manipulation of demographic information. No such interest exists in relation to the demography of any other species of plants or animals. Consequently the ability to gather demographic data in plants in comparison with human populations is dismally low and our knowledge woefully insufficient. Furthermore, until recently there has been little theoretical attention paid to the demographic peculiarities of plants. However, there is evidence of an increasing interest in plant demography (Antonovics 1976; Bradshaw 1972; Harper & White 1971, 1974).

Human demography originated from the necessity of governments to know the size of the population for purposes of military conscription and taxation. Early censuses were very crude head counts. With time, data on age, sex and property were added. Eventually it was realized that knowledge of age-specific birth and death rates and of present population sizes could be used to forecast future population sizes. In human demography the basic statistical unit is the individual, with the family as the secondary unit (Spiegelman 1968). Demographic statistics are gathered in human populations on certain key life history events such as births, marriages, reproduction and death. This emphasis has been adopted in principle also by plant demographers, although there may be reasons to modify it, as we will see.

The principal objectives of plant demography are twofold. First, agriculturalists are interested in obtaining information on optimal sowing densities of crops; effective economical life span of perennial plants such as fruit trees; optimal harvest age of timber crops; and life history characteristics of noxious weeds. Because of the economic importance of agriculture, much demographic data have been collected by agronomists and foresters with a very immediate and practical objective and is often presented in ways that make the data unsuitable for any other purpose (Chapter 7).

A second objective of plant demography and the one with which this book is primarily concerned is to gather demographic data in order to understand the dynamics of the population and the action of natural selection.

Darwin, according to his own account (Darwin 1887), came upon the principle of natural selection as a result of reading Malthus's *Essay on Population*. The concept of natural selection is basically a demographic concept (Harper 1967). However, for a variety of reasons, plant geneticists, ecologists and evolutionists (including Darwin) have preferred to infer the action of natural selection from the existence of presumably adapted characters rather than measuring natural selection directly in natural and artificial populations. The evolutionary and genetic literature holds many more cases of circumstantial evidence for natural selection than of direct proof (Dobzhansky 1970; Grant 1963; Mayr 1963; Stebbins 1950). The time has come for a more rigorous approach.

Modern population biology is in the process of unifying the theoretical knowledge of population genetics whose principal concern has been with changes in gene frequency, with theoretical knowledge of population ecology concerned with changes in numbers of individuals in a population and the forces that control them. Although some progress is being made in developing an integrated theory (Karlin & Nevo 1976; Solbrig *et al.* 1979), there still exists a dearth of rigorously obtained data from natural populations. The reasons for this situation are many and will be presented, at least in part, in these pages, but it is important to recognize that observations of, and experimentation with, simultaneous changes of gene frequency and plant number are now possible. In the present chapter a discussion of how such studies should be structured and the problems involved, both theoretical and practical is presented. This will be followed in subsequent chapters by detailed discussion of some major aspects of this problem.

LIFE HISTORY STRATEGY AND FITNESS

Central to a quantification of demographic variables is a need to comprehend the life cycle of a plant. Unravelling the various aspects of the life cycle is crucial to an understanding of the dynamics of the population and the community. It is also essential for approaching questions such as: What determines germination times, seedling survival, flowering and fruiting times? How does a plant partition its energy between vegetative and reproductive growth? What is the meaning of genetic variation in populations and of variation in breeding systems? What is the effectiveness of vegetative propagation *v.* sexual reproduction? Why is there so much variation in seed size, seed number and dispersal potential between species? Although animal ecologists have been concerned for a long time with the details of the life cycle of fauna, few attempts have been made to understand the adaptive aspects of the life cycle of plants. Probably this is due to the technical details involved in accurately estimating seed production and following their fate over and under the soil surface, as well as monitoring seedling establishment and seedling and adult mortality.

The growing realization that adaptation involves not only biochemical, physiological and morphological characteristics, but also and principally the life history parameters, has led to the concept of a life history 'strategy'. The life cycle of organisms, like any other complex phenotypic characteristic, represents a series of compromises to a set of physical and biological selective forces. The components of the life cycle, such as timing of germination, seedling and adult survival, age of flowering and flower and seed numbers, constitute a life history strategy, implying a set of adaptive responses accumulated over evolutionary time, without any teleological implication (Wilbur *et al.* 1974).

The minimum components of a life history strategy in plants are as follows:

1   soil-seed pool, seedling and adult mortality;
2   age of first reproduction;
3   reproductive life span;
4   fertility, that is, proportion of individuals reproducing at a particular time;
5   fecundity, including number of seeds, which in turn depends upon number of flowers produced and rates of pollination;
6   fecundity-age regression; and
7   reproductive effort, which is the allocation of resources to any reproductive activity as opposed to resources allocated to growth, tissue maintenance and defence against predators.

To assess the adaptive value of the life history characteristics of a plant genetic and environmental aspects have to be considered simultaneously.

PROBLEMS IN MEASURING FITNESS

It is customary to ascribe non-random changes in the genotypic composition of a population to the action of natural selection. Such a conclusion can only be justified as a working hypothesis and does not constitute proof of selection (Prout 1965, 1969). More questionable still is the practice to assign adaptive value to a character (or phenotype) that is changing in frequency, without ascertaining the reasons for its increase (or decrease). In effect, not all genetic changes are the result of viability differences between phenotypes (and underlying genotypes) but they can be the result of differences in fertility, or differential success in mating or gametic competition. A classical example is furnished by the short-tail character (*t*-allele) in mice (Lewontin & Dunn

1960), where the increase in the frequency of the character is due to gametic selection.

To rigorously measure the action of natural selection it is necessary to establish

1   the phenotypic and genotypic composition of the population of zygotes at the very moment of fertilization, prior to any selection; and

2   again just after reproduction when selection has been completed; as well as

3   the contributions of each phenotype (and underlying genotype) to the offspring generation.

This involves measuring the gamete production of each phenotype and underlying genotype, including both ovules and pollen output (whether produced by the same individual or by different individuals), and the success of these gametes in reproduction. Once the differential mortality and differential reproduction of each phenotype and underlying genotype in the population has been measured, and only then, is it possible to arrive at an understanding of the dynamics of the population and the causes of success or failure of given phenotypes.

There are a number of very serious problems confronting anybody who attempts such a massive undertaking. Although plants do not present the problems of capture, mark and recapture that haunt animal demographers, observations on plant survival, flower and seed production and germination and seedling survival are extremely laborious and time consuming. Another problem is that a considerable proportion of the plant is below ground where it is not visible. This includes not only roots, but also seeds. Differential seed mortality is a very important component of selection (Chapter 6). Another very serious problem is the fact that most species of plants have very long life spans, often exceeding the life span of the investigator, not to mention that of the average research project. This difficulty can be overcome by concentrating on annual or short-lived species, but it is difficult to draw generalizations based exclusively on studies of short-lived species. Another serious problem is the fact that determining the genetic composition in plants is usually not possible in a non-destructive way. Biochemical markers (allozymes), which are the preferred characters used today for determining the genotypic composition of a plant, require destructive sampling of at least portions of the plant. When phenotypic markers are used whose inheritance pattern is known, such as flower colour, or chlorophyll-mutants (Apirion & Zohary 1961), they are expressed usually in either the juvenile or the mature state, but not in both. Furthermore, zygotes are essentially impossible to sample because of their size and the fact that they are enclosed in the ovary. Finally, although plants are easy to mark and observe (at least their aerial portions), it is practically impossible to do it without disturbing the population. Consequently the investigator can easily become an unwitting selective agent.

Therefore, the plant demographer can never obtain a perfect understanding of the dynamics of the population. However, by devising efficient sampling methods that take into account the problems just enumerated, and by the use of appropriate statistical techniques, it is possible to acquire a fairly good and rigorous understanding of population dynamics (Prout 1969; Workman & Jain 1966).

The conventional models of population genetics are predicated on the basis of a number of unrealistic simplifying assumptions. So, for example, they usually deal with one or two loci with two alleles; time is measured in generations; breeding is assumed to be random, and selection is measured by the probability of survival from birth to maturity and, sometimes, the average number of offspring per individual, per generation, is also calculated, with the age structure of the population either ignored or not taken into account. Although such models have heuristic value, they are of limited predictive value when dealing with real populations.

Because events in the life cycle that affect overall reproduction occur at different rates at different ages, time cannot be measured in discrete units of one generation. It is not possible even in annual plants, unless it can be established that there is no seed dormancy beyond the resting period between generations. Furthermore, mating in plants is never random and is further complicated by the ability of many angiosperms to self-fertilize and/or multiply vegetatively. Finally, only rarely are phenotypic characters coded by one or two major genes (Chapter 3).

NATURAL SELECTION AND THE LIFE CYCLE

We now take up the question of age-specific survival for various life history stages.

To aid in the presentation of the life cycle it will be divided into five main stages as follows (Figure 1.1):

1 *pre-dispersal phase* (from fertilization to seed release) when the new plant is still dependent on the mother plant;

2 *dispersal phase;*

3 *germination and establishment phase,* from germination to the production of the first pair of true leaves;

4 *adult stage,* including both the pre-reproductive ('juvenile' *sensu* Hickman 1979) and reproductive adult;

5 *flowering, gametogenesis and fertilization.*

PRE-DISPERSAL PHASE

In the seed plants fertilization takes place while the ovule is attached to the megasporophyll. The embryo is nourished by the mother plant which also forms with its own tissues a protective cover (seed coats, fruit). Although the details of embryology, seed and fruit formation, are fairly well known, the

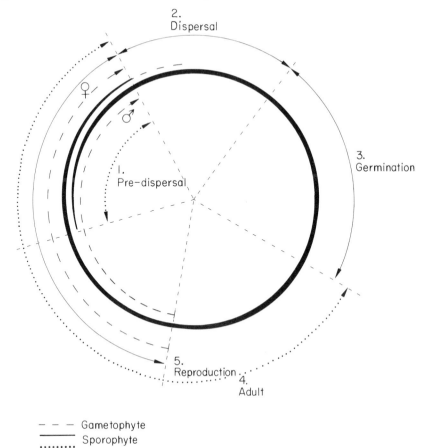

FIGURE I.I. Generalized life cycle of flowering plants. – – – –, Gametophytic phase; ——— and · · · ·, sporophytic phase. Note that in flowering plants gametophyte and sporophyte overlap completely (no free-living gametophyte) as do the adult, reproductive and pre-dispersal phases.

evolutionary meaning of maternal care in plants and its demographic and genetic implications have not been investigated. Curiously little attention has been paid to the trade-off between seed number and individual seed size. Given that the energy (or some mineral factor such as nitrogen) at the disposal of the mother plant is limited, and has to be partitioned between maintenance and reproduction, size of seed will inevitably limit the number of seeds a plant can produce and, inversely, any increase in seed number has to be at the expense of either energy (or whatever factor is limiting) devoted to maternal plant maintenance, or in the size and weight of seeds. Since seeds are the smallest free-living stages in the sporophytic phase of the life cycle, they are most vulnerable. The larger the seed, the higher the probability of successful germination and survival of seedlings (Werner 1977, 1979). Strong correlations between seed size and degree of shading of germination sites (Salisbury 1942), and between seed size and moisture stress (Baker 1972) lend support to the close tie between seed size and physical environment.

Packaging energy in seeds involves a conflict between the advantage a parent gains from producing a large number of seeds and the advantage to the offspring of large seed size with corresponding high individual fitness. C.C. Smith (1975) and Smith & Fretwell (1974) have proposed a graphical method that predicts the optimal solution to this problem under the assumption that selection proceeds in such a way as to strike a compromise between parental and offspring fitness (Figure 1.2). Parent and offspring usually differ in their genetic composition. The relatedness can vary from a theoretical low of 0·5 (the male gamete shares no alleles identical by descent with the ovule) to a high of 1·0 (in apomictic species). The number and relative success of zygotes with different relationships to the mother plant will affect the genetic composition

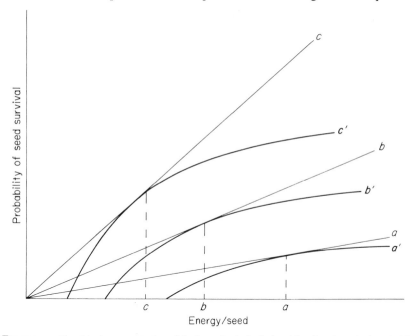

FIGURE 1.2. Graphical representation of the hypothetical relationship of seed survival probability and seed size (energy/seed). Every point along a line (a, b, c) represents the same total number of survivors in the *seed population*. The greater the slope, the larger the total number of offspring surviving. The curves (a¹, b¹, c¹) represent *individual seed* survival probabilities in three different environments. When the seed is very small, its survival probability is so low as to be essentially zero. The relationship is not linear due to size-independent mortality. The point at which the seed populations survival line is tangent to the individual seed survival line represents the seed size where total population survival probability corresponds to maximum individual seed survival for the given environment. It can be equated with the point a compromise which inclusive fitness is maximized. In harsher environments (reduced probability of survival of a seed given size) compromise seed size is larger (after C.C. Smith, 1975).

of the population. When seeds are genetically identical to the mother plant the production of many seeds with low fitness, or of a few seeds with high individual fitness, will not affect inclusive fitness (and genetic structure of the population), provided that the final contribution of reproducing offspring is held constant. But, when each seed in an ovary is genetically unlike, competition within the zygote population will favour seeds that are capable of extracting a larger proportion of the maternal resources (and thereby increasing their own probability of success) at the expense of total number of seeds, setting up a possible conflict between parental advantage (many seeds) and offspring (large seeds). The effects of such conflict should be most readily observed in outcrossing species with multiseeded fruits. The relationship between breeding system and number and size of seed has not been studied but two observations are relevant in this connection. One is the tendency for hybrid seed to be larger than selfed-seed; the other, the phylogenetic trend for a shift from multi-ovulated flowers to uni-ovulated flowers. Any genetic factors that increase competition between zygotes for maternal resources must be transmitted via pollen.

A serious problem with this model is that the majority of the genes in the genotype may be identical by descent in all the individuals of a species. In effect, if we judge by the phenotype, the similarities are much greater than the differences within a species as compared with other species. Consequently, the few genes that are polymorphic would play a very important role in the supposed competition between parent and offspring and between sibs. However, the few direct measures of population heterozygosity (Hamrick 1979; Selander 1976) indicate between 30 and 50 per cent of heterozygosity in populations. If this figure is correct it could be supportive of the model of Smith just described (see also Chapter 3). However, more recent evidence (Finerty & Johnson 1979) questions the high heterozygosity values that have been reported.

DISPERSAL PHASE

Seeds are dispersed both in space and in time. Dispersal in space involves transport by some outside agent, either a physical factor such as wind or water currents, or an animal (mostly mammals and birds). Dispersal in time refers to the fact that once a seed is buried in the soil it may stay in a dormant condition for a considerable length of time.

Dispersal is a process of searching for the most favourable point in space and time for germination. But, because dispersal in plants is a passive process, the seed can only control its movement in a very limited sense, through morphologies of fruits and seeds that increase the probabilities of being moved by wind or water, of adhering to animal coats or of being eaten (Burrows 1973, 1975). Given the low probability that any individual seed has of being dispersed successfully it is not surprising to find that seed production by plants is very high, on the order of 10^3–10^7 seeds/plant/generation, of which, at best, only a very few will become established. With such numbers the selection of favourable seed mutations, or the elimination of undesirable characteristics may proceed quite rapidly.

Solbrig & Cantino (1975) investigated seed production from fertilization to seed dispersal in a group of related species of the genus *Prosopis* in Argentina and the United States. These species of trees produce large seeds in multiseeded pods (legumes), which are produced by multiflowered spicate inflorescences (Simpson, Neff & Moldenke 1977).

Each of the three species studied *(P. chilensis, P. flexuosa, P. velutina)* produced on average between 220 and 240 flowers per inflorescence, of which no more than 2 flowers developed into mature fruits, many inflorescences producing no fruit at all. In *P. flexuosa* only 26 flowers in 10 000 initiates a fruit, and only 7 in 10 000 reaches maturity (Table 1.1). Mature seeds are

TABLE 1.1. Number of flowers, incipient fruits and mature fruits in three species of *Prosopis* (from Solbrig & Cantino, 1975)

Species	No. trees	No. flowers	Mature fruits	
			No.	%
P. flexuosa	10	56 000	40	0·7
P. chilensis	10	46 000	4	0·09
P. velutina	10	111 461	73	0·7

attacked by bruchid beetles. These insects lay their eggs on the fruit surface, the larvae penetrating the seed and developing in them, with total destruction of the seed. From 1 to 25 per cent of the seeds are damaged while still on the tree or shortly after dispersal. As fruits lie on the ground they continue to be reinfected (Figure 1.3). Only seeds in fruits that are removed by mammals (that eat the fruits and discharge the hard seeds in their faeces), or get buried, can escape destruction. Seeds have a varying percentage of germination (which is usually over 50 per cent) after passing through the intestinal track of mammals (Table 1.2). Consequently, it takes a minimum of 100 000 flowers to produce one germinating seed, whose chances of survival in turn are very low! To increase the probability of seed dispersal by mammals (and avoid bruchid predation) these plants produce seeds in fruits with a semi-fleshy, sweet and nutritious mesocarp with a 13·91 per cent protein content (Catlin 1925), a non-trivial cost to these inhabitants of desert and semi-desert areas (see Chapter 8 for further examples of seed predation).

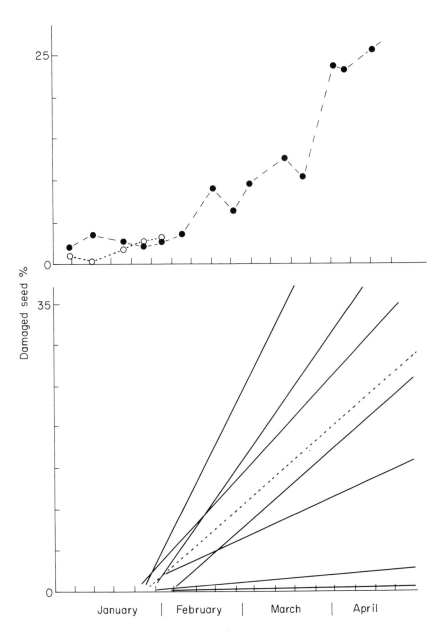

FIGURE 1.3. Percentage of damaged seeds in relation to time in *Prosopis flexuosa*. Upper graph: all ten trees together; lower graph regressions for the logarithmic phase of seed damage for individual trees (*r* varies from 0·96 to 0·50); note the great variation between individual trees. Broken line represents average for all trees (from Solbrig & Cantino 1975).

TABLE 1.2.Germination percentage of three species of *Prosopis* (from Solbrig & Cantino 1975)

Species	No. seeds	Average germination
P. flexuosa	263	54.6 ± 7.9
P. chilensis	2 301	41.3 ± 6.3
P. velutina	90	82.2

GERMINATION AND ESTABLISHMENT PHASE

Germination is one of the two most hazardous stages of the life cycle of the plant, that is some of the highest mortality rates in the life cycle occur at this stage (the other hazardous stage being dispersal). It appears that the two major sources of seedling mortality are water stress (Cook 1979; Mack 1976; Sharitz & McCormick 1973) and herbivore damage (Christensen & Müller 1975). Consequently, we expect seedlings to possess the ability to germinate when these risks are lowest. This may involve germinating only in certain times of the year, or in response to unusually favourable circumstances or germinating when density and potential adult competition is lowest. Baskin & Baskin (1972, 1973, 1975), Beatley (1967), King (1975), Linhardt (1976), Miles (1972, 1973), Platt (1976), Sagar & Harper (1960) and others have documented some of the many adaptations shown by seeds of various species in different environments. It already has been mentioned that survivorship probability is directly related to seed size and weight (Werner 1979). On the other hand, in seasonal environments seed germination is staggered, which was shown by Cohen (1966) to be the optimal strategy to maximize seedling survival in an unpredictable environment.

The negative effect of competition by conspecific adults on survival is shown dramatically in a field experiment with species of *Viola blanda* (Solbrig, unpublished). The number of seedlings germinating in two adjacent 1 × 1 m quadrats and their survival was recorded over the growing season (Figure 1.4). The vegetation of the sample quadrats was a dense mat of *V. blanda*. One of the quadrats was left undisturbed; in the other all adults plants and their underground parts were removed in the autumn of 1976. The number of seedlings emerging in the two plots during 1977 was similar. However, survival in the undisturbed plot was very low (two seedlings remained alive by the autumn of 1978), while survival in the disturbed plot at the end of the season was better than 50 per cent. These observations show very dramatically that seedlings of *V. blanda* normally cannot compete with adults and that their survival is dependent on adult mortality. Seed and seedling characteristics and their relation to a plant's life history strategies are discussed further in Chapter 6.

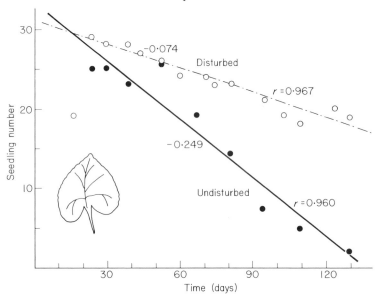

FIGURE 1.4. Seedling survival of *Viola blanda* over time for two adjacent 1 m² plots. ●, Undisturbed population; ○, population where all adults had been removed.

ADULT PHASE

Plants grow by repetition of identical or similar parts, such as buds, leaves or root segments. As individuals increase in size some parts, both aerial and underground, are lost as a result of predation or senescence. Also, different individuals increasingly compete with each other for light, water and nutrients. As individual plants increase in size by accretion of parts, the smaller (weaker?) ones are slowly eliminated by intra- and inter-specific competition, following certain mathematical relationships between size and number (Chapter 2). Demographic data, in the form of life tables, provide important means of quantifying the impact of biotic and abiotic stress factors (Cook 1979).

Mortality patterns can influence the demographic structure of a population as proposed by MacArthur (1962) and MacArthur & Wilson (1967). According to the theory of *r* and *K* selection (MacArthur & Wilson 1967) genotypes that allocate a greater proportion of their resources to reproductive activity (*r* strategists) will be favoured in environments with high density independent mortality, while genotypes that allocate a low fraction of resources to reproduction in favour of vegetative growth are favoured in crowded environments where most mortality is density dependent. Experimental proof is provided by populations of dandelions.

Three populations of the common dandelion *(Taraxacum officinale)* were investigated by Solbrig & Simpson (1974, 1977) in respect to the proportion of carbon allocated to reproductive and vegetative functions. The populations were from:

1 a highly disturbed lawn;
2 a less-disturbed lawn; and
3 an early successional site.

The three sampling sites were separated by less than 500 m from each other. Four biotypes as revealed by their isozyme patterns were discovered. Two of them, named *A* and *D*, accounted for approximately 70 per cent of all plants. *A* was the majority biotype in the disturbed sites 1 and 2, while *D* was the majority biotype in the relatively undisturbed site 3. Experiments and observations show that the *D* biotype is a better competitor than the *A* biotype, but that it sets fewer seeds than *A* under the same environmental conditions (Table 1.3). From these studies it was concluded that the observed

TABLE 1.3. Average number of heads per plant produced by three dandelion biotypes when grown under standard conditions (from Solbrig & Simpson 1974)

	Biotype			
Site of origin	*A*	*B*	*C*	*D*
1	3·6	2·3	1·5	*
2	2·6	2·1	1·9	*
3	3·8	2·3	0·5	1·2

* There were no plants of *D* from site 1. The one plant from site 2 did not bloom.

polymorphism is a direct adaptation to differences in disturbance between the three sites and to accompanying mortality due to the man-made disturbance. The *A* biotype, which has the high *r* value, is favoured in the highly disturbed sites but the *D* biotype, with a greater *K* value, is found in the less-disturbed sites.

The two biotypes (*A* and *D*) were then grown separately and in mixture in garden plots for 4 years. Some of the plots were subjected to artificial disturbance by defoliation or removal of the entire plant at midsummer for the first 2 years. After 4 years biotype *D* accounted for approximately 85 per cent of the plants and 91 per cent of the biomass in the mixtures on the undisturbed plots, but only 7–9 per cent of the plants and 3–7 per cent of the biomass on the disturbed plots.

However, the selective effects of competition alone cannot explain all the diversity of the life history strategies in adult plants (Stearns 1977). Predation and environmental uncertainty have to be taken into consideration also (Wilbur, Tinkle & Collins 1974).

Predation may result in high adult mortality and in such a situation it may be advantageous for the plant to bloom early and produce a large number of seeds. However, the plant may instead evolve toxic compounds that deter herbivores. The energetic cost of producing anti-herbivore substances will result in less competitive plants or in a lower seed crop.

Environmental uncertainty is more difficult to assess. A uniform predictable environment results in a uniform death rate over time. Whether this favours investment in vegetative functions to reduce mortality further (by increasing competitive ability) or in high seed production to reduce the losses due to mortality probably depends on other factors, such as environmental harshness, predation and competition. Unpredictable environments, on the other hand, will favour opportunistic strategies, i.e. maximizing survival during favourable periods but producing enough of a seed crop to overcome unfavourable periods and wide dispersal if uncertainty is patchy.

In a study of seven species of milkweeds, *Asclepias* spp., in Michigan, Wilbur (1976) attempted to partition the relative effect of competition, predation and environmental uncertainty on carbon allocation to vegetative and reproductive functions. The seven species are long-lived perennials which share pollinators and possess identical mechanisms of dispersal. Seed production in each species was interpreted in terms of

1   competition (measured by the proportion of non-flowering plants and by the density of competitors);

2   predation (measured by the frequency of plants damaged by predators); and

3   environmental uncertainty (measured by mortality rates).

The life history strategies that evolved in response to these factors, and the general characteristics of the species were as follows.

A. exaltata is a widespread species found in moist woodlands. It showed a very high proportion of plants visibly damaged by herbivores. So did *A. viridiflora,* a species widespread in old fields. Both species also had a high frequency of flowering individuals, few pods per umbel and a low number of seeds per pod. Both species can sustain a great amount of damage to their aerial parts without suffering mortality because of a large root storage capacity. In addition, *A. viridiflora* is also found as a colonist in disturbed areas. In this environment the flowering season of *A. viridiflora* is long and a large number of seeds are produced. Plants had a high annual survival rate but only about 20 per cent of the plants that flowered escaped herbivore damage and were able to produce fruits. Predation is probably the major selective factor in these two species, with competition as the secondary agent (Table 1.4).

A. purpurascens is very similar to *A. exaltata* in morphology and preferred habitat, but differs from this species in producing more seeds per pod, having a lower herbivore load and a higher frequency of non-flowering plants.

TABLE 1.4. Indices of competition predation and mortality for seven species of *Asclepias* (after Wilbur 1976)

	Plants not flowering		Plants damaged by herbivores		Annual mortality rate of adults	
	%	Sample size	%	Sample size	%	Sample size
A. exaltata	53·7	54	67·4	310	13·1	396
A. incarnata	57·2	140	16·4	140	22·6	159
A. purpurascens	78·3	23	23·3	30	23·5	17
A. syriaca			37·2	484		
A. tuberosa	61·1	144	4·9	144	5·7	455
A. verticillata	48·6	72	24·0	121	13·6	88
A. viridiflora	55·0	129	6·9*	115	14·5	193

* Usually higher in other sites (H. Wilbur pers. comm.)

Environmental uncertainty plays a slightly more important role in the evolution of life history strategy in this species, while competition plays a lower role, as compared to *A. exaltata* and *A. viridiflora*.

A. incarnata is a species of marsh and pond edges, while *A. verticillata* is found in dry hillsides. Both species show a low incidence of herbivore damage, a relatively high mortality rate but an intermediate frequency of non-flowering individuals. Both species have a large number of umbels per stem although *A. incarnata* possesses determinate growth and a short flowering season, whereas *A. verticillata* has a long flowering season and growth is indeterminate. The environments of these species are more uncertain (less predictable) than the three previous species and this factor is considered the most important determinant of the life history (Table 1.4).

A. tuberosa is a common and widespread weed in old fields. It is similar to the preceding pair as far as herbivore damage, but produces fewer seeds per pod and has an extremely low annual mortality rate. This species also has a low probability of successful germination and seedling establishment, probably due to competition and the uncertainty of the physical environment. Finally, *A. syriaca,* a ubiquitous weed of open upland habitats, is the only species with vegetative reproduction, which permits it to persist very successfully once established. It is a copious seed producer, with an intermediate level of predation. The life history strategies of these last two species are presumably influenced most by environmental uncertainty and can best be characterized as 'opportunistic' (Table 1.4).

Thus, the life history strategy in these species cannot be explained taking only competition (and density independent mortality) into account. Other axes such as predation and environmental uncertainty have to be considered. Whether these three parameters suffice in all cases must await further investigation.

FLOWERING, GAMETOGENESIS AND FERTILIZATION

The last demographic stage to consider is fertility. A number of investigators have demonstrated large genetic variances for characters correlated with fertility, such as ovule and pollen number, degree of self-fertilization, seed production, etc. (Allard, Jain & Workman 1968; Hamrick & Allard 1975). However, so far genotypic specific fertility schedules for plant species have not been published. The reasons for this situation are many.

To construct a genotypic specific fertility schedule the investigator needs to determine

1   the genotypic composition of the population under study;
2   the output and genic make-up of the gametes (both pollen and ovules) for each genotype;
3   the mating success and breeding patterns of each kind of gamete; and finally
4   the output and genetic composition of the zygotes and seeds produced by each reproducing adult.

Thus, in plant demography to estimate fertility selection we must simultaneously account for gamete production, gametic selection and mating. Such a schedule is close to impossible. However, by sampling adult plants together with their seeds, the relevant parameters can be estimated (Christiansen & Frydenberg 1973, 1976; Clegg 1979). Recently (Mulcahy 1975), some more direct methods to measure gametic selection have been devised, which can be used to test some of the estimates obtained in the mother–offspring analysis.

The objective of the mother–offspring analysis is to ascribe gene frequency changes that may exist between the maternal and offspring population to one or more of three major possible causes:

1   *sexual selection,* or differential success of different genotypes at mating;
2   *fecundity selection,* or differential zygote production by different matings; and
3   *gametic selection,* or distorted segregation of heterozygotes.

By assaying gene frequencies in mothers and their offspring in a cohort, it is possible to evaluate statistically the contribution of these three components of selection (Table 1.5). Christiansen & Frydenberg (1973) discuss the appropriate statistical tests.

Clegg (1979) and Clegg, Kahler & Allard (1978) have presented a model (Table 1.6) for estimating fertility selection by means of a mother–offspring analysis. The model was tested in experiments conducted with artificial populations of barley. Mother–offspring data of this type are easy to obtain in species where the seeds remain attached to the maternal plant. The breeding system of the species must be known in order to use this model. There is also the further assumption of a uniform distribution of pollen types across the maternal populations. In their study with barley, Clegg *et al.* (1978) calculated fertility selection by harvesting bulk seed in each generation, germinating the

TABLE I.5. Analysis of selection components at one locus with two alleles assuming equal fecundity of the mother–offspring combination.

	Null hypothesis	Possible effects that may lead to rejection of null hypothesis
1	Half the offspring from A_1A_2 plants is heterozygous	Gametic selection in A_1A_2 plants
2	Frequency of transmitted male gametes (pollen) is independent of genotype	Non-random mating in the breeding population and plant (female) specific selection of male (pollen) gametes
3	Frequency of transmitted pollen (male gametes) equals the gene frequency in the population	Differential pollen (male gametes) mating success and gametic selection in pollen
4	Gene frequency in 1-year old seedlings same as in seed population	Selection against specific allele, or gene dependent germination
5	Genotypic frequency in 1-year old seedling same as in seed population	Selection against specific genotypes or genotype dependent germination
6	Gene frequency in adult population same as in 1-year old seedlings	Zygotic selection in adult plants

TABLE I.6. Census points and notation for experimental design to measure viability and fertility components of selection in monoecious annual plant species. The quantities f_{ij} and a_{ij} denote relative genotypic frequencies among the population at zygotic and adult stages. The numbers of progeny of genotype kl observed for the ijth maternal genotype are denoted n_{kl}^{ij} where $i, j, k, l = 1, 2, \ldots$ n (from Clegg, in preparation)

	Census stage		Progeny genotype		
Genotype	Zygotic	Adult	A_iA_i	A_iA_j	Zygotic
A_iA_i	f_{ii}	a_{ii}	n_{ii}^{iii}	n_{ij}^{ii}	f'_{ii}
A_iA_j	$2f_{ij}$	$2a_{ij}$	n_{ii}^{ij}	n_{ij}^{ij}	$2f'_{ij}$
Sample size	N_f	N_a			N'_f

seed in the laboratory under optimal conditions and assaying the gene frequency of selected loci by electrophoresis 7 days after germination. Estimates of gene frequencies in the ovule population (Table 1.7) indicate that fertility selection has a large and important influence on gene frequency distribution in the barley populations tested.

TABLE 1.7. Genotypic frequencies, relative viabilities with standard errors in parentheses, and χ^2 statistics computed from likelihood ratio tests for generations 19 and 28 at locus EC in barley CCV (from Clegg, Kahler & Allard, 1978)

	Generation					
	19			28		
Genotype	Zygotic f_{ij}	Adult a_{ij}	Viability v_{ij}	Zygotic f_{ij}	Adult a_{ij}	Viability v_{ij}
C_1C_1	0·661	0·532	1·00	0·519	0·476	1·00
C_1C_2	0·0	0·001	—	0·009	0·004	0·46 (0·27)
C_1C_3	0·002	0·005	3·43 (2·80)	0·005	0·004	0·76 (0·49)
C_2C_2	0·058	0·096	2·07 (0·35)	0·205	0·256	1·36 (0·14)
C_2C_3	0·002	0·0	0·0	0·002	0·0	0·0
C_3C_3	0·277	0·366	1·64 (0·14)	0·259	0·260	1·09 (0·10)
Sample size	1007	1098		1102	1052	
$\chi^2_{(5)}$		43·76			13·50	

APPROACHES TO THE STUDY OF POPULATION DYNAMICS

In the previous pages some of the complexities of the life cycle of plants and the dynamics of populations have been described. It was shown that even at this level the complexities are enormous, so that by necessity some simplifications will be inevitable. But, if the deduction of general principles rather than the detailed description of the system is the ultimate objective, such simplification, especially if it disposes of unimportant detail, is beneficial rather than burdensome.

In analysing and modelling complex systems, the crucial decision lies in the choice of relevant variables. Three principal approaches to the study of plant populations can be identified based on the principal variable studied:

1 an energetic approach based on the study of the capture, transformation and use of energy;

2 a genetical approach that studies the change of gene frequency in the population; and

3 a demographic approach based on the changes over time in numbers of individuals or parts of individuals in the population.

Plants are, in principle, well suited for a study of population dynamics based on energy flows. Energy intake is a function of a few easily measureable independent variables (light, CO_2, water, temperature) and, given these, photosynthesis is theoretically predictable (Tenhunen, Yocum & Gates 1976; Tenhunen *et al.* 1977); standing biomass is easy to ascertain and so are predation and maintenance losses (McCree 1970; Penning de Vries 1975). However, so far the technical problems of accurately measuring photosynthesis in the field have proven to be large and their solution costly so that no precise, integrated study (or model) of population dynamics based on energy flow is available. Nevertheless, this approach is very promising (Mooney 1976). The advantage of energy as the principal variable in a dynamic model of a population of plants is that it provides a common currency to assess benefits and costs. However, some studies (Givnish 1979; Medina 1971; Mooney & Gulmon 1979) seem to indicate that energy may not be the only basic metabolic currency but that the flow of materials (especially nitrogen) may be equally important.

The advantage of an approach based on gene frequency changes throughout the life cycle is that it integrates ecological and genetical theory. Furthermore, introducing genetical parameters may help explain behavioural dissimilarities between individuals in the population. Also, if the object is to understand evolution by natural selection, genetical information is essential. The problems with a precise genetical model are even more difficult than with an energetic model. Gene frequencies (at least for a few random loci) are not that difficult to obtain since the introduction of the isozyme technique. The interpretation of genetic variation for reasons to be discussed in Chapter 3 is, however, so difficult as to render isozyme data unusable at present other than as genetical markers. Nevertheless, to fully understand plant population dynamics, genetic information is essential.

Finally, population dynamics can be modelled using number of individuals as the relevant variable. This approach has the advantage of apparent simplicity since individuals should be easy to identify and count and, furthermore, the bulk of ecological theory is based on such an approach. However, plants are not as well suited for such an approach as it may appear at first sight because variable and indeterminate growth often invalidates life table data based on age, and vegetative propagation creates two types of physiologically independent individuals: those that are genetically identical (ramets) and those that are genetically unrelated (genets), which can be identified only through genetical analysis (Solbrig 1972).

Consequently, a minimal approach to model plant populations will require an understanding of ramet dynamics and the genetical structure of the population throughout the life cycle. In the following chapters the dynamics of population numbers in adults (Chapter 2) and seeds (Chapter 6) and the problems of vegetative propagation (Chapter 5) are reviewed as well as the genetic constraints in populations (Chapter 3) and the further complications introduced by breeding systems (Chapter 4). These concepts are then considered together in discussing two ecological systems: agricultural systems (Chapter 7), which are the simplest and perhaps best understood and tropical systems (Chapter 8) which probably are the most complex and least understood. The emphasis is on a balanced approach that combines theory with empirical information. Although no exhaustive review of the literature was undertaken, the more than 750 entries provide the most exhaustive reference list in plant demography and related subjects published to date.

Chapter 2
Demographic Factors in Populations of Plants

JAMES WHITE*

Adequate syntheses of demographic phenomena in plants have been rare until recently and there is much further work to be done. This is not to say that demographic data are unavailable for plants, for there is a plethora of them (Harper & White 1974), but they have been collected by plant scientists with a variety of interests in a wide range of species (mostly crop plants). The absence of an awareness that demographic phenomena have great significance for understanding evolutionary processes has hitherto been a conspicuous feature of all but a handful of plant biologists. The publication of 'Charles Darwin's Natural Selection' (Stauffer 1975) and the collected papers of Charles Darwin (Barret 1977) show clearly how Darwin sought plant demographic data on cultivated plants from horticulturalists as eagerly as information about breeds of pigeons from pigeon fanciers. Allen (1977) has documented this more fully. But there were no subsequent attempts to integrate such work into evolutionary theory until recently (Harper 1977). Perhaps the fragmentation of the plant sciences—forestry, agronomy, horticulture—alongside 'traditional' botany, each with its lines of demarcation as to what species were suitable for investigation, has hindered a more holistic appreciation of the value of diverse demographic data for interpreting evolutionary dynamics. The work of John Harper and his school has been conspicuous in its attempts to collate the varied approaches of plant scientists to population biology, though it seems there is still little attempt by applied botanists to relate their work to any central core of demographic theory. While the abundance of plant demographic data for crops has become more evident recently (e.g. Harper & White 1974), these can only be used (as Darwin did) to complement studies in natural environments with less human intervention, since they are commonly defective in important observations or measurements which an evolutionary biologist would consider essential (Chapter 7).

   In this chapter, I shall draw heavily on the work of applied botanists, especially foresters, to demonstrate the value of their work to demographic theory.

* Many of the data reviewed in this paper were compiled at Harvard Forest, Harvard University, where the hospitality of the staff and the facilities of the library helped greatly.

DEMOGRAPHY OF THE INDIVIDUAL PLANT

Some of the earliest accounts of plant demography have less to say about numbers of individual plants than about the multiplicity of parts within a single plant. Plants are plastic in their growth form with few exceptions, developing branches, leaves and often inflorescences to a variable degree depending on environmental circumstances. The demographic consequences of growth forms in plants have been little considered, though Bell (1976) has attempted preliminary analyses of dispersal of herbaceous perennials as a function of their rhizome geometry. And yet the notion of the plant as a population of parts is quite old.

One of the earliest expressions is that of Richard Bradley. 'The twigs and branches of trees are really so many plants growing upon one another; for as they all proceed from buds, we may thence infer that the buds they came from did in every respect perform the office of a seed' (Bradley 1721: 41). The analogy between seed and lateral bud persisted in the work of Gaertner (1788) and Goethe (1790) through to Erasmus Darwin (1800: 1) '. . . . every bud of a tree is an individual vegetable being; a tree therefore is a family or swarm of individual plants, like the polypus, with its growing young out of its sides or like the branching cells of the coral-insect'. This was presumably the source of his grandson's similar analogy (Darwin 1839).

A thorough treatment of the notion of the plant individual as a population was provided by Braun, who discussed the subject more philosophically than empirically. However, he came to the conclusion that 'as far as we are justified in speaking of vegetable individuality at all, we must hold fast to the individuality of the shoot: the shoot is the morphological vegetable individual' (Braun 1853).

There are several discussions of this theme since, which will be reviewed elsewhere (White, in 1979). Thus, for over 200 years one of the recurrent approaches to plant demography has been through the plasticity of the individual plant. It seems appropriate too that this should be a focal point of demographic studies on organisms in which number and size are often closely related (White & Harper 1970).

It seems best to adopt a pragmatic approach to the basic unit of plant structure which is to be counted: buds or leaves on small annual plants, buds or shoots on herbaceous perennials, shoots or leaf clusters on shrubs and trees: there are examples of all in a widely scattered literature. The unit should at least be countable practically and it should be one which can reasonably be expected to vary with clear environmental factors such as energy or nutrient levels, or with density stress from separate genetic individuals or with interference from other, similar, units. For example, the dynamics of leaves have been enumerated for subterranean clover (Stern 1965), for flax (Bazzaz & Harper 1977) and for a *Tilia* shrub (Miyaji & Tagawa 1973), the dynamics of tillers observed in grasses (Langer 1956) and bamboos (Kadambi 1949) and the numbers of branches counted on red maple trees (Wilson 1966).

While the growth plan of plants is to a large degree genetically determined, their realized size and shape depends greatly on environmentally induced rates of formation and loss of vegetative parts. This has been clearly documented by foresters (Millington & Chaney 1973). Temperate trees (such as birches, poplars, oaks, maples, beeches) rarely show more than seven orders of branching (that is, clear articulations between successively larger branches from primary shoots to the main trunk) (Büsgen & Münch 1929). A specimen of *Taxus baccata* (yew) with nine orders of branching merited a report (Löhr 1965). There are trees and shrubs which may show more than nine distinct articulations, for example *Rhus typhina* (sumac) and other trees conforming to Hallé & Oldeman's (1970) model of Leeuwenberg. Divaricate shrubs in New Zealand (Went 1971) may also show exceptions to what is otherwise a general rule.

Shedding of twigs and branches by trees may be caused ultimately by wind action, as one may casually observe, but their death is normally due to competitive interactions within the crown. 'The growing shoot behaves towards the other parts in an egotistical fashion . . . a large number of shoots cannot develop at all, others remain very small . . . without this correlation every crown would be a witchbroom. Only the never-ending struggle of all shoots against all others produces the shape of the trees which is not only physiologically harmonious, but which we consider aesthetically pleasing' (Münch 1938, translation provided by M. H. Zimmerman). Tree shape is clearly an outcome of demographic events, but these have never been studied in detail.

Mar-Möller, Müller & Nielsen (1954) sought to measure the loss of branches in *Fagus sylvatica* (beech) by collecting fallen branches and twigs: their results are expressed in volume and biomass terms, not numbers. They record an annual loss of branches in 50-year old beech trees as 12·8 per cent of the annual dry matter increment in stem and branches. It is also clear from the forestry literature that trees commonly have specialized absciss joints at which twigs and branches are shed (Millington & Chaney 1973). Much remains to be known about the demographic phenomena which underlie tree shape.

Interactions between plants modify the balance between the 'birth' and 'death' processes involved in plant development. Grazing animals usually harvest some but not all of the units, leaving shoots or branches to develop further. In short the demography of subunits (or modules) is the starting point for the demography of genetically distinct plants (genets) (Harper & White 1974, White 1979).

DEMOGRAPHY OF MODULES

A basic unit chosen for counting may be simple or complex. Counting leaves or buds may be straightforward with flax or clover, but very difficult with trees (though entomologists seem to have done even this, e.g. Allen 1976). A fruitful approach is offered through the concept of plant architecture introduced by Hallé & Oldeman (1970). The visible morphological expression of the genetic blueprint of a plant at any one time is its *architecture,* whereas the growth programme which determines the successive architectural phases is called the *architectural model,* or plan of growth. To date twenty-four such models have been recognized for trees in tropical forests; many temperate trees seem to fit one or other of a relatively small group of them (Hallé, Oldeman & Tomlinson, 1978). The models are defined *a priori* on criteria of primary (extension) growth: life span of meristems, differentiation of vegetative meristems, plagiotropy or orthotropy of axes, episodic or continuous growth of axes and chronology of branch development. Many of the models seem applicable to the growth patterns of some herbs and shrubs in temperate zones (e.g. Temple 1977) although of course other systems of growth-form classification have been applied repeatedly to these plants (e.g. Mueller-Dombois & Ellenberg 1974). One of the striking innovations of Hallé & Oldeman's system is its emphasis on structural features with distinct architectural analogies. Their concept of 'article' (French), best rendered in English as module, is essential to an understanding of all branched structures in trees—an indefinite succession of modules constitutes the architectural models they define. Each module may vary from model to model, just as the assembly of different modular components may lead to distinct architectural styles in buildings. The module is formally defined as an axis whose meristem creates all the differentiated structures of a shoot from inception to flowering. While a module may itself be plastic, it does seem to offer some compromise between the total enumeration of buds or leaves on plants and a complete disregard for the plant as a population. Generally, the module becomes smaller in size as the plant grows older and accumulates modules (see later).

In some architectural models the modules are more easily recognized at a glance than in others. For example the model of Tomlinson is defined as being basally branched, with all axes orthotropic and equivalent and with inflorescences commonly terminal on each axis. It is the model of many palms and in a miniaturized form of many grasses. (The models are named after botanists rather than specific plants to avoid a too rigorous typological interpretation which might be associated with a particular named plant species). A demographic study of tillers is obviously straightforward and there are several examples available (e.g. Garwood 1969; Langer 1956, 1963; Langer, Ryle & Jewiss 1964). Agronomists have commonly used tiller counts as an indicator of plant size and vitality. This is not to say that tillers are invariably uniform in size, no more than are the leaves they carry. But heuristically they prove to have more value in assessing the performance of grass species than simple counts of clumps alone (Hyder 1972).

A striking example of the insights which a demographic approach at the

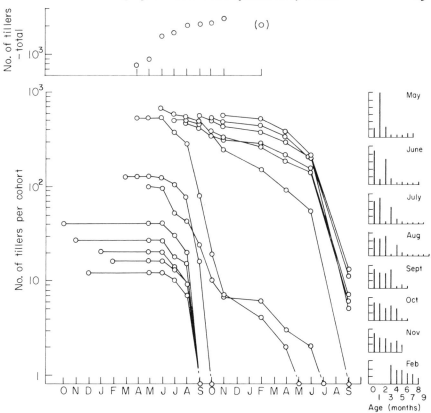

FIGURE 2.1. Survivorship curves of successive monthly cohorts of tillers of *Phleum pratense*. The numbers in each cohort are summed over forty plants each grown from seed in soil in separate 20-cm pots. The abscissa indicates the months of the year. The total number of living tillers as the population reaches an asymptote is also shown, together with the age structure at successive monthly intervals during this period. Each division on the ordinates of the age structure graphs is 10 per cent (Drawn from data in Table 8 of Langer (1956).)

level of the module provides may be gleaned from the study of Langer (1956) on *Phleum pratense* illustrated in Figure 2.1. Langer followed at monthly intervals the development and fate of tillers on each of forty plants grown singly in soil in 20- cm diameter pots. It is clear graphically (his data were presented in tabular form) that successive cohorts of tillers (counted over all forty separate plants) behave in a remarkably similar fashion, showing a typical Deevey Type 1 curve (Deevey 1947). Mortality of tillers is slight until an inflorescence is formed and each tiller dies after a single reproductive effort. Tillers formed in June do not flower until the following year. Some tillers formed in April and May may flower shortly afterwards, but otherwise wait until the following year: this gives a two-phase survivorship curve to these

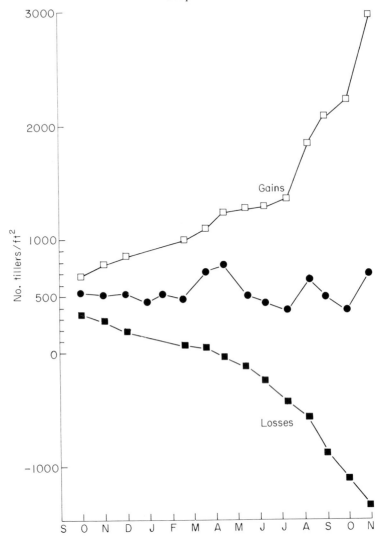

FIGURE 2.2. Population dynamics of tillers of *Phleum pratense* in permanent quadrats (0·09 m²) on 5-year-old planted swards. ●, Standing population at each census interval; □, cumulative gains of tillers; ■, cumulative deaths of tillers. (Drawn from data contained in Figure 5 and 8 of Garwood (1969).)

cohorts. If the age structure of the tillers is calculated for any given month an interesting trend is noted. Until August the total number of tillers had increased rapidly to an asymptotic number about 2 000 and there is consequently a changing age structure until then, as shown. Afterwards there is tentative evidence that as the total number remains more or less the same, death being only slightly overbalanced by recruitment of new cohorts, a stable age distribution of tillers begins to emerge. Unfortunately no further cohorts were observed after November so that one cannot decipher the age structure later on, although existing cohorts continued to be observed until the following September. Obviously through the winter months there will be some recruitment so that the age of the population will change, as partly outlined in the data shown for February (of course the percentages shown for ages 3–8 months are overestimates since the numbers of 0–2-month tillers are not known). It seems that the age structure may be cyclically stable depending on the environment, with a lesser risk of death in winter than in summer. A more extended age structure may develop in the autumn to spring months than during the summer when only five or six monthly cohorts contribute to the population. In any event, there is a demographic pattern in the dynamics of tillers, which has hitherto been unappreciated and further work of this sort over longer periods is desirable.

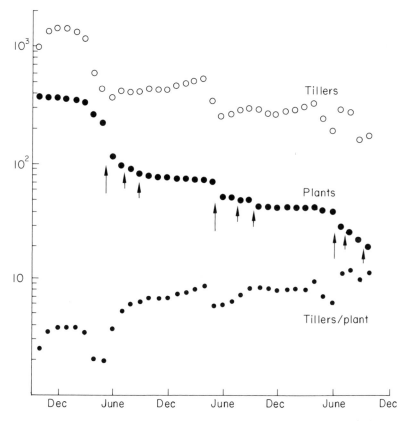

FIGURE 2.3. Changes in numbers of seed-sown plants, tillers and tillers per plant of *Phleum pratense* grown in sward conditions for 3 years. Arrows indicate cutting times thrice yearly, at flowering time and later. Ordinate shows numbers ft^{-2} (1 ft^2 = 0·09 m^2). (Drawn from data contained in Figures 1, 2 and 3 of Langer *et al.* (1964).)

28 *Chapter 2*

Some other aspects of grass tiller dynamics have been demonstrated by Kays who showed that tillers of *Lolium perenne* decrease in number (self thin) under density stress according to the $\frac{3}{2}$ thinning rule (Kays & Harper 1974). Figure 2.2 shows an example of tiller dynamics of *Phleum pratense* grown in sward conditions (Garwood 1969). Permanent quadrats were established and all live tillers were marked at monthly intervals and their development recorded. Notwithstanding seasonal fluctuations the numbers remain within certain limits, a result of a dynamic balance between formation and death of tillers. Similar results have been obtained for distinct genetic individuals of *Plantago* spp. (Sagar 1970) and *Ranunculus* spp. (Sarukhán & Harper 1973).

The relationship between genet and module dynamics may be important in elucidating the changing genetic structure of plant populations under natural conditions. The studies of range scientists on grasses may be a fruitful source of such information. In an experiment on *Phleum pratense,* Langer *et al.* (1964) neatly documented such a relationship, depicted in Figure 2.3. *Phleum* seedlings were established in large (*c.* 1 m³) containers at a density of *c.* 400 plants ft $^{-2}$ (4 300 m $^{-2}$) and allowed to grow for 3 years. Seedlings and tillers were marked and monitored every month in two small permanent quadrats totalling 1 ft² (= 0·09 m²) in area in the centre of each container. The swards were cut at time of flowering in midsummer each year, followed by two further cuttings before winter. The number of genetically distinct plants declined twentyfold in 3 years. After an early increase in numbers the tillers settled down to a population density fluctuating between *c.* 200–500 ft $^{-2}$. The number of tillers per genetic individual surviving increased from 2 to 10. It is worth noting the stepwise survivorship curve of tillers and genetic individuals at or before the time of cutting. This seems to be a further example of a phenomenon to which Sarukhán drew attention for *Ranunculus*: mortality is generally most severe at a time of rapid growth in biomass in spring and early summer and not during wintertime (Sarukhán & Harper 1973).

Closely related to the modular structure of grasses is that of bamboos. They are, however, defined somewhat differently as the model of McClure (Hallé, Oldeman & Tomlinson, 1978), since the vegetative axes are mixed, that is differentiated into orthotropic and plagiotropic axes, with the basal branches producing new (usually subterranean) trunks. However, the general similarity of bamboo clumps to grass tussocks is readily apparent and there is little difficulty in recognizing and counting modules. There are many accounts of shoot growth and shoot dynamics in various bamboo species, but few with adequate primary data to reconstruct some basic demographic generalities. The work of Kadambi (1949) is, however, especially useful since it contains pages of tabulated raw data on the numbers of *Dendrocalamus strictus* culms. There are doubtless multitudes of similar data available from experimental stations, temperate and tropical, for trees, range grasses, bamboos and other species, but they are rarely published *in extenso* and rarely interpreted by those who collect them in a manner that makes sense demographically. Often the

TABLE 2.1. Age structure of culms of *Dendrocalamus strictus*. The age distribution of individuals at the time of first observation (shown in parentheses) is unknown. Cohorts suffer no mortality until they are 6 years old. The area of plot 3 is 1 800 m^2, the area of plot 1 is not given. Recalculated and rearranged from data in Kadambi (1949).

Plot No.	No. culms (clumps)	Time period (years)	Age distribution (years)						
			0–1	1–2	2–3	3–4	4–5	5–6	>6
1	245(17)	T_1	14	(231)					
	270	T_2	25	14	(231)				
	295	T_3	25	25	14	(231)			
	321	T_4	31	25	25	14	(226)		
	338	T_5	48	31	25	25	14	(195)	
	379	T_6	66	48	31	25	25	14	170
	422	T_7	56	66	48	31	25	25	171
	428	T_8	32	56	66	48	31	25	170
	437	T_9	44	32	56	66	48	31	160
	%	T_6	17·4	12·7	8·2	6·6	6·6	3·7	45
		T_7	13·3	15·7	11·3	7·3	5·9	5·9	40·5
		T_8	7·5	13·1	15·4	11·2	7·2	5·8	39·8
		T_9	10·1	7·3	12·8	15·1	11·0	7·1	36·6
		Mean T_6–T_9	12·2	12·2	11·9	10·0	7·7	5·6	40·5
3	787(57)	T_1	56	(731)					
	919	T_2	132	56	(731)				
	1 067	T_3	180	132	56	(699)			
	1 239	T_4	292	180	132	56	(569)		
	1 304	T_5	181	292	180	132	56	(463)	
	1 472	T_6	168	181	292	180	132	56	(463)
	1 502	T_7	160	168	181	292	180	132	389
	%	T_6	11·4	12·3	19·8	12·3	9·0	3·8	31·4
		T_7	10·6	11·2	12·0	19·4	12·0	8·7	25·8
		Mean T_6–T_7	11·0	11·8	15·9	15·8	10·5	6·2	28·6

observations span several decades, seemingly maintained long after their instigators have retired or died. They are important sources, almost impossible to duplicate given modern research fashions, and should be more thoroughly investigated.

A few of Kadambi's observations have been recalculated and tabulated in Table 2.1. Data from two of his five experimental plots are shown. At the beginning of his observations he counted the number of established clumps and culms on the plots (the size of plot 1 is not given, plot 3 was 1 800 m^2). At successive yearly intervals for up to 9 years all new culms recruited to the population were recorded. Older culms were not individually observed and since their ages were not known, one cannot fully reconstitute the survivorship curves for the annual cohorts. However, he did observe that 'all culms survive

Chapter 2

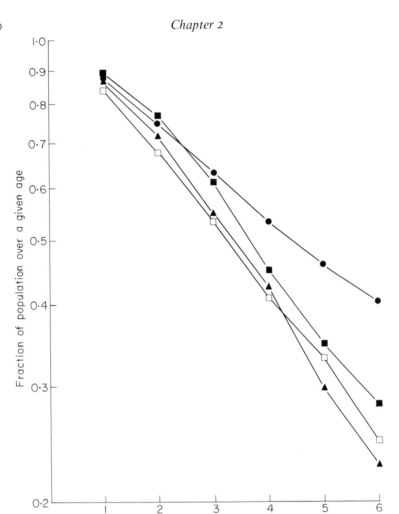

FIGURE 2.4. The fraction of the population of culms of *Dendrocalamus strictus* over a given age on four separate experimental plots. ●, Plot 1, mean of 4 years' observations; ■, plot 3, mean of 2 years' observations; □, plot 4, mean of 3 years' observations; ▲, plot 5, mean of 2 years' observations, based on calculations similar to those shown in Table 2.1 for plots 1 and 3. Average number of culms per plot from which these data were derived was 417 (area unknown), 1 487 (per 1 800 m²), 404 (per 1 600 m²) and 691 (per 1 600 m²), respectively. The fraction of culms over 6 years old (varying from 23 to 40 per cent) cannot be further partitioned from the data available. Drawn from data in Kadambi (1949), rearranged and recalculated.

almost without exception for about 5 to 6 years after which they begin to die . . . the oldest culms on record are 10 years old'. This enables one to reconstruct the data matrices shown in Table 2.1, in which successive annual cohorts move diagonally until at least year 6 from formation, then in year 7 to

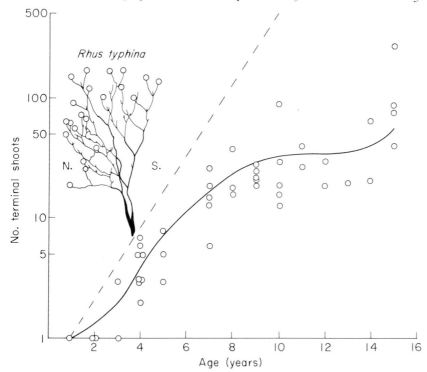

FIGURE 2.5. Relation between age and the number of terminal shoots on 'female' plants of *Rhus typhina*. A diagrammatic outline of part of one tree 4–5 m high is shown. Following flowering or shoot tip abortion on each axis usually two further shoots are formed. Branch formation is more abundant and more vigorous on the south-facing aspect of each bush; o inflorescences. The hatched line indicates the (geometric) increase on an assumption of regular branching by which each axis gives rise successively to two further axes. (The assistace of P. B. Tomlinson and P. Kenmore in gathering these data is acknowledged.)

join the totals of uncertain age dating from the beginning of the experiment; these culms of uncertain age had been declining in numbers slowly (not exponentially). At year 6 on both plots, and in successive years as far as the data permit, there is a full row of culm numbers whose ages are known accurately, except of course for those over 6 years old. The relative percentages in each age class in any given year from year 6 onwards remains remarkably constant, an elegant example of a stable age distribution of modules.

A summary of the age distribution on four separate plots is given in Figure 2.4, where the regular age structure for at least the first 6 years' growth of populations of culms may be seen. There is a clear relationship between the numbers of culms in successive cohorts in the populations: over 80 per cent of the culms are more than 1 year old, 23–40 per cent are more than 6 years old. Extrapolation of the curves to a base line of 0·05 or 5 per cent of the

population would give ages of 12–13 years for three plots and 19 years for plot
1. With the exception of plot 1 this is reasonably close to the maximum
longevity recorded for a culm of *Dendrocalamus strictus*. Unfortunately, there
seems to be no data on survivorship of these culms beyond 6 years of age, but
the shape is likely to be similar to that shown in Figure 2.1 for *Phleum* tillers, a
Deevey Type 1 curve.

Kadambi also studied the effects of various thinning treatments on his
experimental plots and while the details are too lengthy to record here, regular
patterns of cohort emergence and relative numbers between cohorts become
established in these plots also.

Other architectural models *sensu* Hallé & Oldeman may be more complex
and the modular construction less easy to discern. The model of Leeuwenberg
is defined as a branched trunk, with equivalent orthotropic axes; each axis
gives rise to more than one successive axes following the abortion or flowering
of its apical meristem. *Rhus typhina, Syringa vulgaris* and *Ricinus communis*
are typical temperate examples, although large forest trees in the tropics may
be constructed in this manner. No example of branch demography has been
studied but Figure 2.5 shows a preliminary study of branch number on *Rhus
typhina* (sumac) of various ages at Harvard Forest, Massachusetts. The
number of terminal shoots is clearly asymptotic rather than geometrically
increasing. Shaded lower branches develop slowly, branch once, remain
vegetative, or die. Competition between the modular segments of the tree is
evident.

It is also clear from this example that modules while they may be morphologically and physiologically determinate (see Nozeran, Bancilhon & Neville
1971 for a more general treatment of this phenomenon) may differ in size. The
use of the concept of modular structure while lending further refinement to the
demography of the individual plant still does not offer a 'basic unit' of plant
construction which is not plastic. Number and size still interact at this level of
demographic analysis in plants, just as at the genet level. Generally, as module
number increases during plant development, size decreases. Oldeman (1974)
referring to the 'réitération' or recapitulation of the architectural model of
trees, makes the point clear, qualitatively: '. . . *Par analogie physiognomique,
nous appellerons* réitération arbustive *l'édification d'un modèle réitéré de
dimensions modestes. Quand le modèle est encore plus petit, nous parlons d'
arbrisseaux et, parallèlement, d'une réitération en arbrisseau. Enfin, lorsque le
modèle est réitéré partiellement ou en miniature, ou les deux, nous décrivons ce
processus comme une réitération herbacée. . . Les vagues de réitération ont
successivement un caractère arborescent, arbustif, frutescent et herbacé; la
rapidité de la succession de telles vagues dépend du biotope'*. As branch order
number increases from the trunk (order 1) to shoots with only primary
extension growth (generally order 6–8), shoot dimensions become smaller.
The actual numbers of branches of successive orders is highly determinate
$(R_B = N_{(n)}/N_{(n+1)}$, where R_B is the branching or bifurcation ratio, n is order

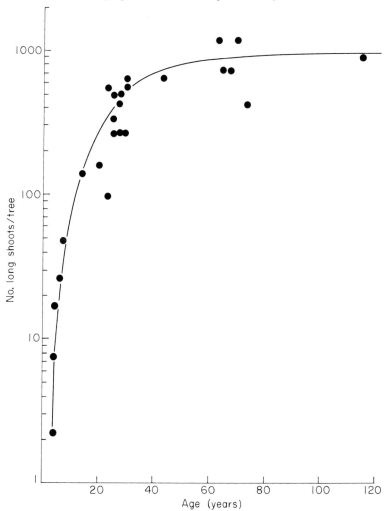

FIGURE 2.6. The number of long shoots on *Acer rubrum* (red maple) trees of various ages. Redrawn from data derived from Figures 5 and 10 of Wilson (1966).

number, N is the number of segments in a particular order), as a variety of workers have shown following an application of Horton's theory of channel networks in drainage basins (Barker, Cumming & Horsfield 1973; Holland 1969; Legay 1971;Leopold 1971; MacMahon & Kronauer 1976; Oohata & Shidei 1971; Whitney 1976). But as branch or module number increases, their average size decreases, according to another proportionality, called the diameter ratio by Barker *et al.* (1973) and Thornley (1977), who based their calculations on diameter, and the length ratio by Leopold (1971), who measured lengths of branch segments on *Abies, Pinus, Fraxinus* and *Liriodendron* trees. Notwithstanding this recent work there is still no clear understanding of the demography of branches on trees and of the relationships between numbers and size as the 'architectural blueprint' of the tree manifests itself. A *rapprochement* between tree modellers and tree biologists remains to be made, but the new insights on plant morphology offered by Hallé, Oldeman & Tomlinson (1978) must be a foundation.

The task of counting and measuring branches on trees is tedious, but it seems that it can be achieved as Wilson (1966) showed. Figure 2.6 shows the number of long shoots (shoots which develop more than 2 cm per year and normally bear lateral branches if more than 1 year old) on red maple trees of various ages (determined by ring counts of felled trees). The results mirror those in Figure 2.5 for *Rhus*, showing a trend towards an asymptotic number after about 40 years (notwithstanding the increased variance in the data beyond this age!). *Acer rubrum* is assignable to the model of Rauh of Hallé & Oldeman (1970), a model common in temperate (as in tropical) trees (e.g. species of oak, ash, maple and pine). Its modular structure is quite regular: a monopodial trunk with episodic growth giving rise to whorled or subwhorled tiers of branches; the branches are orthotropic and morphologically identical with the trunk, being monopodial with episodic growth, and giving rise to lateral branches in a whorled or subwhorled arrangement. Inflorescences are always lateral, on branches or on the trunk and have little influence on the vegetative growth pattern.

Flower-Ellis (1971) has examined the demography of *Vaccinium myrtillus* branches in detail. His work is perhaps the most thorough analysis to date on the demography of plant modules, in a shrub of more complex architecture than either *Acer rubrum* or *Rhus typhina*.

The study of plant form has been the subject of numerous investigations which have been inadequately reviewed since Troll's monographs (1937–43). The importance of growth form studies for understanding plant community structure has been repeatedly emphasized by Dansereau (1961, 1971) while both Lems and Meusel have shown their value for interpreting the adaptations of plants to particular environments (e.g. Lems 1960; Meusel & Mörchen 1977). Plant morphology underlies the plant demographic techniques of the Rabotnov–Uranov school of Soviet ecologists (Rabotnov 1969; Uranov 1967; 1968, 1974; see also Harper & White 1974). Recently several studies in Europe and America indicate that new and fruitful links are being made between plant morphology and demography (Bell 1974; Callaghan 1976; Callaghan & Collins 1976; Smith & Palmer 1976). However, the application of demographic concepts to the study of plant form remains rudimentary.

DEMOGRAPHY OF
GENETICALLY DISTINCT PLANTS (GENETS)

Most work on plant demography, whether with crops or wild species, has been concerned with the demography of genetically distinct plants. Our present understanding of this subject has recently been comprehensively reviewed (Harper 1977).

Some of the best documented generalizations of plant demography concern the relationships between density and size of plants under conditions where competition between plants is clearly occurring (a few workers would dispute that competition occurs between plants or minimize its effect, e.g. Went 1973, but this is a minority view). There have been numerous quantitative formulations of these relationships (Willey & Heath 1969), but the approach of Kira and his colleagues (Kira, Ogawa & Sakazaki 1953; Yoda *et al.* 1963) seems clear and straightforward, since they use a minimum of undetermined constants in their equations and eschew the polynomial equations of high order (and biologically unintelligible constants) so beloved of foresters and horticulturalists. The frame of reference for their analyses of competition-density effects may be stated as follows. Over a wide range of planting densities in artificially created stands there is a convergence in time to a similar yield of biomass per unit area, for all stands except those with very low densities. The formulation of this statement was first proposed for herbaceous plants by Kira *et al.* (1953) as follows

$$w = k_i d^{-1}, \text{ or } Y = wd = k_i$$

where w is mean plant weight, d is density, Y is yield per unit area and k_i is a constant. The value of k_i may increase as the stand develops. This result is an outcome of the ability of plants to develop plastically in response to density stress. Analyses of many spacing experiments in agriculture, horticulture and forestry can be shown to conform to this relationship. In terms of the earlier discussion of module demography, one might say that plants develop greater or lesser numbers of modules, or other structural subunits depending on density stress. This has been clearly shown for branches and leaves of *Trifolium subterraneum* (Stern 1965) and for leaves of *Linum usitatissimum* (Bazzaz & Harper 1977). The life expectancy of leaves was found to be greater at high than at low density in *Linum,* and possibly also in *Trifolium* where the data are not explicit enough. Such detailed demographic studies are still very rare.

However, if planting densities are very high—beyond the usual (culturally learned and transmitted) rates of sowing—or if seedling density of natural populations is very high (perhaps as a result of local dispersal around parental plants) then, as the population grows and develops, the plastic response to density stress may be augmented by mortality. In circumstances of high plant density without the intervention of substantial extrinsic density-independent regulatory forces, thinning or density-dependent mortality usually results as the population develops. Intuitively one can appreciate that, while a certain density of seedlings or saplings may be maintained in early life, as individuals grow larger competition for resources may become severe, leading either to intraspecific mortality for the 'weaker' individuals or to growth stagnation.

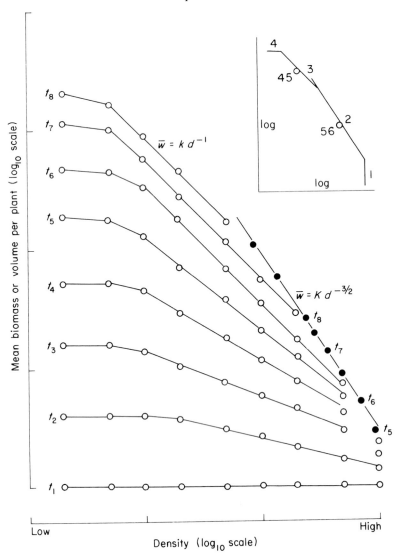

FIGURE 2.7. Generalized graphical scheme for representing competition-density effects, after Kira *et al.* (1953) and Yoda *et al.* (1963). Subscripted *t* indicates successive time periods. The order of thinned stands (•) along the line of −3/2 slope is diagrammatic and does not indicate any fixed or necessary sequence. The inset shows four of the salient feature of the density–yield relationships.

Stand stagnation is reported for forest trees (e.g. for *Pinus contorta*, Smithers 1961) but mortality is the commoner result of high density stress.

The first formal analysis of the behaviour of pure stands undergoing density-dependent mortality or self-thinning seems to be that of Tadaki & Shidei (1959) with more elaborate treatment given in subsequent papers, largely from a forest management viewpoint (Ando 1962; Tadaki 1963, 1964). Extensive treatment of the process was given by Yoda *et al.* (1963) for pure stands of herbaceous weedy species. The relationship between surviving plants and level of performance of the population is expressed by these authors in terms of surviving stand density and mean biomass per plant of the survivors. This may be expressed as:

$$w = Kd^{-\frac{3}{2}} \text{ or } Y = wd = Kd^{-\frac{3}{2}}d = Kd^{-\frac{1}{2}}$$

The symbols are the same as in the previous equations, but K is a different constant. Further examples of this relationship are given in White & Harper (1970). Figure 2.7 gives an example of the graphical conventions involved. Initially the mean plant biomass is independent of density but, as the populations develop (successively subscripted t), plants at the higher densities grow on average more slowly than those at low density. For a certain range of density the interplant stress is absorbed by plasticity and the relation $w = k_t d^{-1}$ develops, with increasing values of k. Overlying this pattern is the 'thinning line' $w = Kd^{-\frac{3}{2}}$ which 'cuts off' as it were the lines of $45°$ slope. Stands at higher densities are constrained by this line earlier than those at lower densities. Subsequently, the stand trajectory in time is along this line. No combinations of density and biomass occur beyond it. Some populations may not be encompassed by this line within their life time. Populations at very low densities may not conform to either relationship (§ 4 of inset). Of course, all possible combinations of density and mean plant biomass are possible within these constraints, but not outside them, so far as is known from extensive analyses of competition and spacing experiments (White, unpublished). It should perhaps be emphasized that the thinning line is a statement of constraint on numbers and size in populations undergoing density-dependent mortality. All possible density-independent effects resulting in very varied number–size relationships may occur within this constraint.

The relationship holds for mixed stands of forest trees taking all species collectively, but not for separate species which behave in a variety of ways depending on their ecological status and tolerance (White, unpublished).

The data available for an analysis of this sort are quite plentiful in the literature of applied plant science, especially in forestry. They are not easy to duplicate without years of measurement and the co-operation of many people. Figure 2.8 shows an example of thinning in even-aged stands of *Pinus ponderosa* taken from the work of Meyer (1938). This line is based on 450 sample plots mostly 400–2 000 m² in size, with some up to 4 000 m², sampled 'by seven or more investigators and their assistants in five [U.S.A.] national-forest

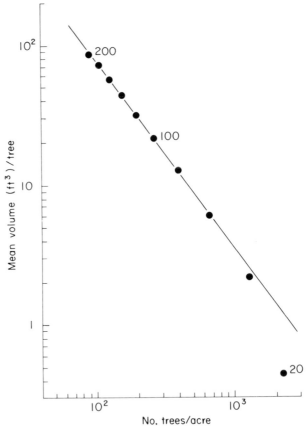

FIGURE 2.8. Self-thinning in even-aged stands of *Pinus ponderosa*. Ages in years are denoted in 20-year intervals. Drawn from data in Meyer (1938). (1 ha = 2·47 acres; 1 m³ = 35·3 ft³.)

regions'; to these were added records from a further 398 plots (Meyer 1938). Such resources have never been available to plant demographers outside national forestry services! Figure 2.8 does not show the thinning trajectory for a single stand through time, but is based on discrete even-aged stands of known ages. As Yoda *et al.* (1963) showed clearly for *Erigeron canadensis*, 'all the stands, growing from widely different initial densities and on soils of different fertility level, actually grew along a single curve given by the $\frac{3}{2}$ power law'. Ponderosa pine grows extensively in even-aged stands and some stands commonly thought to be uneven-aged are composites of even-aged groups (Meyer 1938).

The generality of this relationship between plant density and mean plant biomass is known for about eighty species (White, unpublished). Not only is the slope of the line $(-\frac{3}{2})$ constant, but it seems that the value of K is somewhat constant too, given the range of data surveyed. Table 2.2 gives a

TABLE 2.2. The relationship between mean plant weight w (in g) and density d (plants per m²) for plant populations undergoing density-dependent mortality in pure stands. K and a are constants. The constants were calculated from raw data in the papers cited, except those of White & Harper (1970) and Yoda *et al.* (1963): results in these two sources are either original or recalculated therein from other primary sources not listed here. The number of each species refers to its position in Figure 2.9.

	Species 1–11 herbs; 12–31 trees	$\log w = \log K$ $- a \log d$		Author
		$\log K$	$-a$	
1	*Erigeron canadensis*	4·31	−1·66	Yoda *et al.* (1963)
2	*Plantago asiatica*	3·89	−1·48	Yoda *et al.* (1963)
3	*Medicago sativa*	3·93	−1·42	White & Harper (1970)
4	*Trifolium pratense*	3·86	−1·33	White & Harper (1970)
5	*Triticum* sp.	3·83	−1·39	White & Hárper (1970)
6	*Amaranthus retroflexus*	3·85	−1·48	Yoda *et al.* (1963)
7	*Ambrosia artemisiifolia*	3·66	−1·48	Yoda *et al.* (1963)
8	{ *Brassica napus* *Raphanus sativus* }	4·19	−1·41	White & Harper (1970)
9	*Helianthus annuus*	3·84	−1·33	White & Harper (1970)
10	{ *Carex lacustris* *C. rostrata* }	3·98	−1·46	Bernard & MacDonald (1974) and Gorham & Somers (1973)
11	*Erigeron canadensis*	3·92	−1·51	Yoda *et al.* (1963)
12	*Prunus pensylvanica*	3·85	−1·43	Marks (1974)
13	*Abies sachalinensis*	4·34	−1·54	Yoda *et al.* (1963)
14	*Populus tremuloides*	3·64	−1·33	Pollard (1971, 1972)
15	*Corylus avellana* (coppice)	3·61	−1·30	Jeffers (1956)
16	*Chamaecyparis thyoides*	3·88	−1·49	Korstian & Brush (1931)
17	*Picea mariana*	3·53	−1·48	Hatcher (1963)
18	*Pinus strobus*	3·78	−1·70	Spurr *et al.* (1957)
19	*Populus deltoides*	3·08	−1·80	Williamson (1913)
20	*Pinus contorta*	3·88	−1·40	Smithers (1961)
21	*Alnus rubra*	3·46	−1·54	Smith (1968)
22	*Carya* spp.	3·06	−1·73	Boisen & Newlin (1910)
23	*Populus tremuloides*	3·58	−1·48	Baker (1925)
24	*Castanea dentata*	3·54	−1·63	Frothingham (1912)
25	*Liquidambar styraciflua*	3·65	−1·66	Tepper & Bamford (1960)
26	*Pinus monticola*	4·01	−1·63	Haig (1932)
27	*Tsuga heterophylla*	4·08	−1·51	Meyer (1937)
28	*Quercus* spp.	3·53	−1·65	Frothingham (1912)
29	*Abies concolor*	3·98	−1·57	Schumacher (1926)
30	*Pinus ponderosa*	4·06	−1·33	Meyer (1938)
31	*Pseudotsuga menziesii*	4·00	−1·54	McArdle & Meyer (1930)

Species calculated on the same basis, not shown in Figure 2.9

	Species	$\log K$	$-a$	Author
	Chenopodium album	4·00	−1·38	Yoda *et al.* (1963)
	Fagopyrum esculentum	4·41	−1·48	Yoda *et al.* (1963)
	Betula spp.	3·94	−1·63	Yoda *et al.* (1963)
	Picea rubra	3·85	−1·57	Meyer (1929)
	Quercus spp.	3·86	−1·71	Khil'mi (1957)

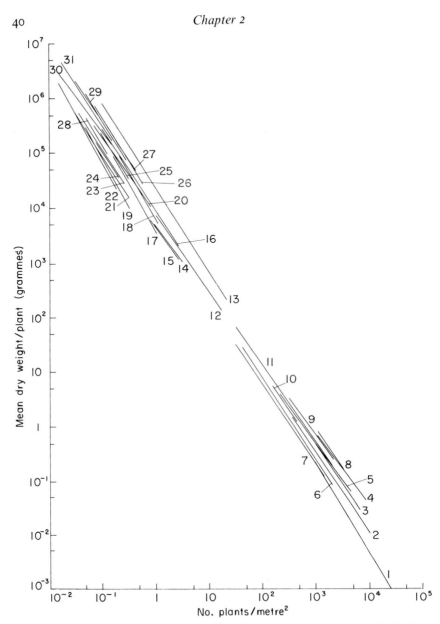

FIGURE 2.9. Self-thinning in herbs and trees. Each line represents the species listed in Table 2.2 and the range over which the observations were made. Derived from sources indicated in Table 2.2.

partial list in which a common data base has been used for all the species (density as plants m^{-2} and biomass as g per plant, shoots only): K is, of course, dependent on the dimensions of these parameters. Data for herbs are straight-forward in this respect, but since foresters usually give tree volume (in cubic or hoppus feet) the appropriate transformation from volume to weight is required. Unfortunately, the volume–weight relationships for trees are quite variable and only approximate values are possible (based on U.S. Dept. Agriculture, *Wood Handbook*, 1940, for weights per cubic foot at 12 per cent moisture content). Notwithstanding such difficulties, all the values of log K shown lie between 3·5 and 4·3, with two exceptions. This is approximately a sixfold difference in linear terms. Figure 2.9 shows most of these species graphically. It seems that the thinning rule (or the '$\frac{3}{2}$ power law' as Yoda *et al.* (1963) referred to it) is one of the more general principles of plant population biology. Only two examples are so far known which lie outside a line given by log $w = 4·4 - 1·5$ log d, *Trifolium subterraneum* with log $K = 4·94$ (Westoby 1976) and *Lolium perenne* with log $K = 4·5$ (Kays and Harper 1974).* However log K is quite sensitive to small changes in the slope which may vary a few degrees so that the interpretation of values of K within an order of magnitude seems unwarranted.

Recently, foresters have become interested in short rotation and intensive culture management for wood fibre production, *Platanus occidentalis* and *Populus* species and cultivars being particularly favoured. High planting density, rapid growth and the inevitable density-dependent mortality have been documented under these conditions, but the combination of numbers and mean plant weight does not exceed the constraint of the thinning line (e.g. results of Ek & Dawson 1976, Harms & Langdon 1976, Saucier, Clark & McAlpine 1972, Wood, Carpenter & Wittwer 1976). Foresters outside Japan have not been aware of this constraint, with very few exceptions (e.g. Drew and Flewelling 1977; Harms & Langdon 1976; T.O. Perry, North Carolina State University, personal communication).

It remains to be seen whether the $\frac{3}{2}$ thinning rule is characteristic primarily of genets such as trees and annual or biennial herbs, to which almost all the examples relate. There are a few cases in which modules may show this pattern of thinning: *Lolium perenne* (Kays & Harper 1974), *Carex* spp. (Bernard & MacDonald 1974; Gorham & Somers 1973), *Corylus avellana* (coppice) (Jeffers 1956). However, one suspects that in long-lived herbaceous perennials population regulation may commonly take place at an earlier stage in shoot ontogenesis: inhibition of lateral bud development may avoid the conse-quences of self-shading and mutual interference between genetically similar shoots. Only injury or fragmentation of rhizomatous structures may allow

* The value of log $K = 5·04$ for *Chenopodium album* in Yoda *et al.* (1963) is based on fresh rather than dry weight, an error in the paper only detected by the deviation of log K from its more typical value (confirmed by T. Kira, pers. comm., Nov. 1977). All the other results in Yoda *et al.* (1963) are for plant dry weight; the corrected value for *Chenopodium* is log $K = 4$.

buds which would otherwise remain dormant to grow as, for example, in *Agropyron repens* (Chancellor 1974) or *Cyperus esculentus* (Thullen & Keeley 1975). Regulation of shoot density in this manner would seem to be especially necessary in conditions of limited energy or nutrient supply and indeed is very obvious in herbaceous perennials of woodland (e.g. Anderson & Loucks 1973; Bell 1974; Holm 1925; Hutchings and Barkham 1976; Struik & Curtis 1962). Species of more open habitats, plentifully supplied with light and nutrients may be expected to allow many more modules to develop, particularly if they can thereby monopolize resources by forming almost monodominant stands e.g. *Carex lacustris* (Bernard & MacDonald 1974), *Glyceria maxima* (Matthews & Westlake 1969), *Phragmites communis* (Haslam 1970). Subsequent death of vegetative modules will almost inevitably occur. However, there is not adequate evidence on this point to warrant more than this tentative speculation, since there are too few pertinent observations on natural populations; the observations on *Lolium perenne* and *Corylus avellana* cited both involved deliberately managed experiments.

However general the thinning line which governs the plant number-plant size relationship, it is still a crude statement of constraint, whose underlying rationale remains elusive. While doubtless based on the geometry of three-dimensional structures packed on a flat surface (White & Harper 1970), it seems to be insensitive to the precise geometry of different species, a conclusion also noted by Westoby (1976) in his analysis of five cultivars of *Trifolium subterraneum*. Perhaps at very high density the gross morphologies of plants are less diverse than when they are grown in isolation. It seems that a more perfect understanding of the subunitary or modular structures and their dynamics may afford clearer insights than we have at present. Attempts to relate thinning to the leaf area index of populations (Westoby 1977; Dewberry 1977) were invalid (White 1977), but doubtless further knowledge of foliage dynamics of individual plants (rather than LAI of stands) will add much to our understanding. Again, it is worth noting, for plants the demography of subunits underlies the demography of discrete individuals.

It is not known how rapidly along the thinning line a stand will develop, although it is a function of soil fertility (White & Harper 1970; Yoda *et al.* 1963). So far no general theory has emerged about how quickly various species of different growth habits or differing ecological strategies move along the thinning trajectory. Nor is it clear just how far a given stand travels along it in time and what its subsequent number–size relationship becomes (although Figure 9 in White & Harper 1970 indicates a theoretical possibility).

The manner in which mortality is induced in populations undergoing self-thinning is probably best understood by examining the pattern of development of large, intermediate and small individuals—the 'hierarchy' of dominant and suppressed plants. The details of this pattern are complex but Figure 2.10 gives a general impression, showing the development of *Pinus ponderosa* in even-aged pure stands over a period from 20 to 160 years of age.

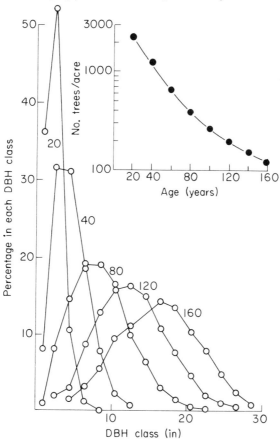

FIGURE 2.10. Diameter distributions in even-aged stands of *Pinus ponderosa*. Age in years is indicated on each curve. The inset shows the decline of numbers in each age class through time. (1 in. = 2·5 cm; 1 ha. = 2·47 acres) Drawn from data in Meyer (1938).

(The thinning line for these populations is shown in Figure 2.8.) The density of plants declines in time, almost exponentially. The frequency distribution of stem diameters *in even-aged stands* shows that the range of diameters increases greatly with age: trees 160 years of age may have stem diameters varying from 12 to 70 cm! Unfortunately data for ages younger than 20 years are not available. It is those trees which are largest in the young stand which persist; the smaller ones die out (Meyer 1938). There is a profusion of forestry evidence to support this view that the smallest trees die (e.g. Lee 1971). The smaller trees are not necessarily lower in the canopy in even-aged monocultures such as this, since stem height is usually maintained at the expense of diameter increment and crown development.

While the frequency distributions of stem diameter remain normal (with

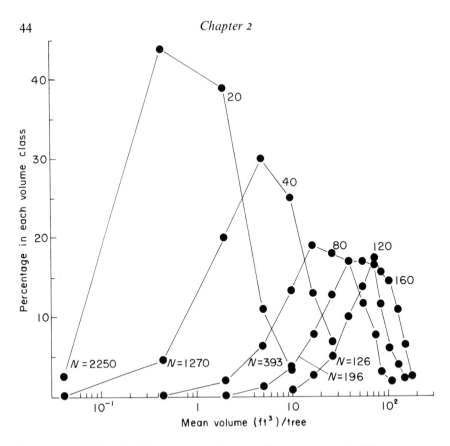

FIGURE 2.11. Volume distributions even-aged stands of in even aged stands of *Pinus ponderosa*. Age in years is indicated on each curve. N=denotes the actual number of stems at each age. Drawn from data in Meyer (1938). ($1 \text{ m}^3 = 35 \cdot 3 \text{ ft}^3$)

increasing kurtosis) through the life of the population, the distribution of tree volume is rather different (Figure 2.11): here the stem diameters shown in Figure 2.10 have been converted to volumes and the percentage in each volume class plotted. Stem volumes do not remain normally distributed but become rapidly skewed to give a few very large individuals. The frequency distribution of stem volumes is virtually log normal. This seems to be a typical result for forest trees though only a few examples have been analysed in this fashion (White, unpublished). Such volume distributions emphasize the dominance–suppression 'hierarchy' which develops early (over a 100-fold difference in size at 20 years) and persists throughout the life of the population (a twentyfold difference at 160 years). Such a volume distribution is to be expected on theoretical grounds (Koch 1966) if diameter distributions are normal, since diameter (D) and volume (V) are related as $V = kD^a$, where k and a are constants. Commonly, a has a value of about $2 \cdot 5$ (White, unpublished).

SIZE STRUCTURE AND AGE STRUCTURE
IN PLANT POPULATIONS

The investigation of module and genet dynamics to this point has stressed the degree to which numbers and size are intertwined in plant population biology. The specification of density without also a statement of size has less meaning in the demography of organisms which may develop plastically (like plants) than in organisms with a more restricted variability in morphological expression (such as higher animals). Indeed, size (or extent of modular development) may also be a more useful criterion than age in plants for determining life expectancy or reproductive schedules, as the Rabotnov–Uranov School has shown. There is abundant evidence that size or vigour are better predictors than age of the onset of reproduction or of repeated reproductive output (Harper & White 1974; Werner 1975b). Perennial herbaceous plants are commonly difficult to age, anyway, unless they accumulate clearly decipherable annual increments (examples given by Harper & White 1971). Most herbaceous plants are not so accommodating to the demographer, however. In some cases perhaps age may be inferred from the size of clonal patches as Oinonen did for a variety of species (see Harper & White 1974); and Whitford (1951), for forest herbs or from a good understanding of the geometry of growth and dispersal (Bell 1974, 1976).

Trees of course offer much more favourable material for determining age structures of populations, at least in temperate and boreal forests, and some use has been made of the almost perfect equivalence of calendar year and growth-ring increment in trees of such forests. However, the equivalence does not hold for tropical or subtropical trees as Tomlinson & Craighead (1972) have shown so clearly. While size and age of trees may be correlated in a very general way, there is no such necessary relationship. Unless there is explicit evidence on this point, the interpretation of size as age may be misleading and lead to simplistic or even inaccurate conclusions.

Nowhere is this more evident than in the plant ecological literature on succession in temperate forests, in which ecological theorists drawing on the abundant and easily derived measurements of tree size (especially stem diameter, allometrically easily related to total volume or biomass) deduce the age structure of tree populations and predict the future course of succession. There are clear examples of this procedure in the works of Daubenmire (1968:105) and Horn (1971:38). Of course some understanding of succession may be gleaned from the distributions of sizes *per se,* especially of local small scale succession as Oliver (1975, 1978, Oliver and Stephens 1977) has carefully documented. But the equivalence of size and age should always be demonstrated rather than assumed because the diameter of a trunk is a function of the growth rate as well as of age. In general it seems that there is a reasonably close relationship between size and age of trees of early successional stages

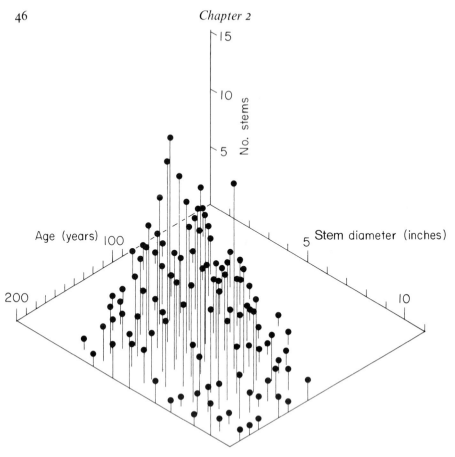

FIGURE 2.12. The relationship between stem diameter at breast height (1·5 m) and age at the root collar for 480 trees of *Picea mariana* at Lake St Pierre, Quebec, based on all stems in ten 100 m² plots. There were few records in the 1–3 in. diameter classes because of scarcity of saplings of seed origin. (1 in. = 2·5 cm.) Drawn from data in Table 2 of Hatcher (1963).

(e.g. Spring *et al.* 1974) or 'intolerant' species, while the relationship is quite poor for tree species of later successional stages or 'tolerant' species (using intolerant and tolerant *sensu* Baker 1949) (Toumey & Korstian 1937). Virgin forests show extreme variability in the tree age–size relationship (Blum 1961; Gates & Nichols 1930; Gibbs 1963; Hough 1932; Morey 1936; Tubbs 1977). There are few studies on the age structure of forest trees to equal those of Hough & Forbes (1943) or Mauve (1931) in which large numbers of trees were accurately aged (7 500 specimens were examined by Hough & Forbes to reconstruct the forest history). Auclair & Cottam (1971) took cores from 854 stems of *Prunus serotina* to determine age distributions (58 per cent of them were essentially even-aged). Managed, second-growth forests may show a good age–size relationship even for late successional species such as *Acer saccharum* (Tubbs 1977), although in the original forest 300-year-old trees ranged from 30 to 80 cm in diameter at breast height (Tubbs 1977).

Figure 2.12 shows details of an analysis by Hatcher (1963) of the age and diameter distribution of 480 stems of *Picea mariana,* sampled in several discrete tracts of forest. There were few records of stems in the 1–3 in. diameter classes because of difficulty in locating spruce saplings of seed origin. Otherwise, the data show clearly the degree of diameter dispersion possible in trees of a given age. If the relation between size and age is so variable for pure stands how much more misleading will it be for mixed stands with more complex dynamics. There seems to be no substitute for age directly observed in plants if age structures are to be discussed accurately. It has commonly been assumed that a reverse J-shaped plot of stem numbers against diameter represents an all-aged rather than even-aged forest, a notion perhaps best popularized by Meyer (1952). He made it clear that such diameter distributions not only characterized virgin forests, but any large forest area ('an entire county or state'). While the latter is probably true, it is not clear whether or not virgin forests have such diameter distributions *and* are all-aged (Jones 1945). Such a curve may be constituted by a series of discrete even-aged classes each showing the normal curve typical of an even-aged stand such as those for *Pinus ponderosa* (Figure 2.10). Obviously the scale of analysis is critical in assessing whether a forest is even-aged or all-aged: on a landscape scale there will almost inevitably be a juxtaposition of even-aged patches of forest to give a reverse J-shaped curve of diameter and age distribution. An example of this is given by Whittaker (1975:8), although the inferences drawn seem scarcely warranted by the age-diameter data actually given in the original source he cites (Miller 1923). It was in this sense that deLiocourt (1898) originally introduced this curve. But it is quite unlikely that the same is true of local-scale forest structure: the reverse J-shaped curve of diameter distribution will almost certainly not imply an all-aged stand, especially in a mixture of species where individual growth rates can vary so much even within essentially even-aged mixtures (Oliver 1975). It seems that attempts to interpret the past or future course of forest succession which rely on age structures derived primarily from diameter distributions must be viewed sceptically. Diameter or size distributions alone may give valuable insights into successional processes if allied to a thorough understanding of the biology of the species involved.

Recently the frequency distribution of stem diameters in forests has been approached more critically and several models have been adduced to summarize the patterns detected (Bailey & Dell 1973; Bliss & Reinker 1964; Goff & West 1975; Hett & Loucks 1976; Leak 1964, 1975), some of which have attempted to relate the diameter distribution to age structure. Based on careful analyses of diameter and age Hett & Loucks (1976) found that a sine wave model appeared to fit best the observed age structure, possibly reflecting cycles in establishment. Goff & West (1975) found that density-diameter distributions of natural old-growth stands of small area or uniform structure were of a rotated sigmoid form. They imply that this model holds also for age structure since 'in general, stem age and diameter were correlated', but their evidence on this point is unfortunately vague since adequate details of number of stems aged or the age–diameter relationship are not presented. Leak (1975) attempted to define the age distributions for four species in a virgin spruce-fir stand in New Hampshire, but the results presented as smoothed curves for 20 or 40 year age groups are less than convincing to support his conclusions.

It seems that age structure analyses of plant populations are still rare (Harper & White (1974) review some other examples), just as examples of plant life tables are uncommon. That is to say, plant demography is still rather rudimentary.

Each individual in the species must leave on the average one offspring each generation if the population is not to go extinct. Since the progeny of a heterozygote segregates homozygous recessives of lower average fitness (the so-called segregational load) the fertility of the heterozygote must be higher than that of the 'normal' homozygote by at least the equivalent of the segregational load. In a two allele system where allele A has a frequency p and allele a a frequency q, and where the relative fitness of the three genotypes is $AA: 1-s$; $Aa: 1$; and $aa: 1-t$, the loss of zygotes ascribable to heterosis is $sp^2 + tq^2$. If we assume that the adaptive values of all homozygotes are equal $(1-s)$, then the loss of zygotes due to a single locus is s/n where n equals the number of alleles participating in heterotic combinations. For N independent loci, the loss equals $(s/n)^N$ (Wallace 1963). Table 3.4 shows the number of fertilized eggs a plant must produce on the average in its lifetime if it is to tolerate heterotic systems involving various kinds of homozygote disadvantages for different number of alleles at each locus and different number of independent heterotic loci. It can be seen that very few lethals $(s=1)$ can be tolerated. If the heterotic advantage on the other hand is very slight $(s=0\cdot01$ or lower) then a relatively large number of heterotic loci can be maintained in

TABLE 3.4. The approximate number of zygotes a hermaphrodite plant must produce in order to maintain heterotic system involving a number of independent gene loci and genes with specific deleterious effects when homozygous (modified after Wallace 1963)

Disadvantage of homozygote	Heterotic alleles	Number of loci			
		1	10	100	1000
1·0	2	2	1024	$1\cdot27^{30}$	10^{300}
	10	2	9·3	$4\cdot9 \times 10^9$	$8\cdot12^{96}$
	20	2	2	169	$1\cdot89^{20}$
0·5	2	2	18	$3\cdot11^{12}$	10^{124}
	10	2	2·87	$3\cdot7 \times 10^4$	$5\cdot72^{45}$
	20	2	2	12·5	$9\cdot89^{10}$
0·1	2	2	2	169	$1\cdot89^{22}$
	10	2	2	7·5	$5\cdot9 \times 10^8$
	20	2	2	2	150
	2	2	2	2	150
0·01	2	1·005	1·05	1·65	150
	10	2	2	2	2
	20	1·005	1·005	1·05	1·65
0·001	2	2	2	2	2
	10	2	2	2	2
	20	2	2	2	2

the population. However, with such slight advantages the probability of losing alleles, as a result of randon events and small population size increases significantly. Nevertheless, some data in the literature indicate that the heterotic contribution of single loci may often be of a low order (Sprague 1969).

Hageman *et al.* (1967) conducted a number of studies to try to relate heterosis in corn to the behaviour of enzymes. They studied two different enzyme systems. One was concerned with energy transfer in germinating seeds and seedlings and included triosephosphate dehydrogenase (TPD), aldolase (ALD) and glucose-6-phosphate dehydrogenase. In the second group only one enzyme, nitrate reductase (NR) which is involved in nitrogen metabolism was studied. Four inbred lines (WF9, M14, Hy2, Oh7) were selected without prior knowledge of the level of enzyme activity. Two single-cross hybrids (WF9 × M14 and Hy2 × Oh7) were made. Heterosis in seedling growth was observed in both crosses. In the WF9 × M14 hybrids heterosis was also observed for TPD and possibly also ALD (as measured by increased enzymatic activity in the hybrid over the expected mid point between parental values), but only intermediate activity for G-6-PD. In the other cross (Hy2 × Oh7) no heterosis was observed at the enzyme level since all three systems were intermediate in activity between parental values. For NR activity results were qualitatively similar, heterosis being detected at the enzyme level only in a minority of crosses (Beevers *et al.* 1964). Sarkissian and Skrivaslava (1967) have also detected heterosis at the molecular level using mitochondrial complementation techniques. These and other studies point out that, although some increased enzyme activity can be detected in hybrids, that it is not universal, and that when detected it is not present in all enzymes, nor are the effects so far found very dramatic.

There are numerous studies that show that heterosis (i.e. hybrid superiority) is a real and important phenomenon (Allard 1961). But fitness increases can arise as a result of more efficient energy transfer mechanism in the plant, higher assimilation rates, better utilization of photosynthate, better or more efficient translocation of minerals and/or photosynthate, increased water use efficiency, changes in morphology, hormonal changes, etc. Each of these factors is a complex process under the control of more than one gene system. It is therefore not surprising that investigations of single loci yield no or only weak support for the observed heterosis at the whole organism level. The superior fitness of hybrids probably results from the addition of many small effects and/or the interaction of various enzymes and biochemical pathways, what is usually called *coadaptation.*

COADAPTATION AND STORAGE OF
GENETIC VARIATION

In the preceding pages it has been shown that in most populations there is ample genetic variation at all stages in the life cycle. The genetic mechanism underlying this variation is however quite diverse. In some cases (albino seedlings) it consists of a single locus with two alleles; in other cases (e.g. plant colour in maize) more than one gene and many alleles are involved. In the latter cases the genes may all have complementary effects, or they may oppose each other; each gene may have effects similar to the other in the system, or some may have stronger effects than others. Individual genes may have measurable effects on the character and may be classed as major or Mendelian genes, the effect may be barely perceptible and the factor will be classed as a polygene. When alleles are identified by their isozyme phenotypes a tremendous amount of variation is revealed. So far the record is held by the xanthine dehydrogenase locus in *Drosophila persimilis* where Coyne (1976) detected no less than twenty-three isozyme variants among sixty females that had been made artificially homozygous in the laboratory. It must be remembered however that using standard techniques over thirty incompatibility alleles (*S*-alleles) were identified in *Petunia* (Whitehouse 1950). In effect the variation identified at specific gene loci is so great that many geneticists are questioning its selective value and increasingly are referring to the variation as selectively neutral (Nei 1973).

As Lewontin (1974) has pointed out, the variation that has been revealed is so large that it can not be explained by models derived from one or two locus genetics. But as has been emphasized in previous pages these models are already invalidated by what we know regarding gene action and the biochemistry of development. But if genes that determine the fitness of an organism do not act independently from one another, then the fitness of a given gene is not an absolute but a relative value, since it depends on the particular genetic background, and should be represented not by a single number (w or s) but by a population mean and a variance. To the extent that the genes in the population are selected so that individuals will on the average have well adapted genotypes, the genes of the gene pool of the population are coadapted (DeBenedictis 1978). Coadaptation of this sort can be demonstrated experimentally if the average viability breaks down following interpopulational hybridization. Such experiments have been carried out successfully with *Drosophila* (Brncic 1954; Dobzhansky 1970; King 1955; Wallace 1955). The existence of integrated genomes adapted to special environments has been shown in *Avena* (Allard 1975; Hamrick and Allard, 1972) and earlier in a variety of species by Clausen, Keck & Hiesey (1940). Rollins & Solbrig (1973) showed the existence of integrated genotypes in natural and artificial populations of *Lesquerella* through a correlation analysis of twenty-three morphological characters.

More recently Allard and coworkers have conducted a number of experiments with barley and oats. In one of these studies (Clegg, Allard & Kohler 1972) four esterase loci (designated arbitrarily as *A, B, C, D)* were simultaneously scored at three time intervals in two artificial populations. These populations called Composite Cross II and Composite Cross V were obtained by pair-wise hybridization of twenty-eight, and thirty different varieties of barley from all over the world. Individual plants were scored for their genotypes at the four esterase loci and four-locus joint gametic frequencies were estimated from the genotypic data over generations (Table 3.5). As the populations evolved, striking correlations developed between loci in allelic states including correlations between non-linked loci. In addition, the same pair of four-locus complementary gametic types came to predominate in both populations. These experiments indicate that selection was operating so as to favour sets of alleles that presumably are coadapted, that is, interact functionally in such a manner that they increase the fitness of the individuals that carry them.

The concept of the genome as a coadapted system of genes and hence the unit of selection is becoming accepted generally (Allard 1975; Grant 1975; Lewontin 1974; Mayr 1963). What is still missing but is being developed (Karlin & Nevo 1976) is a well-developed body of rigorous, predictive, mathematical theory. Such a body will, however, require novel approaches (Levins 1975).

TABLE 3.5. Relative gametic frequencies and their relative deviations (in parentheses) from products of allelic frequencies for one early, one intermediate and one late generation of CCV (from Clegg, Allard & Kahler 1972)

	Generation		
Gamete	5	17	26
1111	0·125 (−0·012)	0·173 (+0·004)	0·126 (−0·027)
1112	0·185 (+0·009)	0·122 (−0·025)	0·052 (−0·005)
1121	0·078 (+0·004)	0·081 (−0·025)	0·110 (−0·066)
1122	0·078 (−0·016)	0·119 (+0·027)	0·067 (+0·001)
1211	0·003 (−0·009)	0·000 (−0·016)	0·005 (−0·049)
1212	0·004 (−0·011)	0·000 (−0·014)	0·000 (−0·020)
1221	0·025 (+0·019)	0·054 (+0·044)	0·229 (+0·167)
1222	0·025 (+0·016)	0·013 (+0·004)	0·022 (−0·001)
2111	0·131 (+0·006)	0·171 (+0·040)	0·175 (+0·078)
2112	0·180 (+0·019)	0·129 (+0·015)	0·106 (+0·070)
2121	0·071 (+0·004)	0·052 (−0·030)	0·081 (−0·031)
2122	0·072 (−0·014)	0·065 (−0·006)	0·022 (−0·019)
2211	0·005 (−0·006)	0·004 (−0·008)	0·000 (−0·034)
2212	0·017 (+0·004)	0·015 (+0·004)	0·001 (−0·012)
2221	0·000 (−0·006)	0·000 (−0·007)	0·002 (−0·038)
2222	0·000 (−0·007)	0·000 (−0·007)	0·001 (−0·014)
N^*	1 452	2 443	3 049

* N denotes the number of zygotes observed.

GENETIC STRUCTURE AND DEMOGRAPHY

If the genome is the unit of selection and if the effect of a given allele on fitness depends on the genetic background of the population, the contributions of alleles to fitness are variables and must be represented in terms of means and variances. But the fitness of an organism in turn is not only a function of its own genetic composition but is also a function of the environment in which it lives. The environment as here defined is both the physical and the biological environment (i.e. the set of competitors and predators). So, for example, Langridge (1962) found an enhancement of heterosis (as measured by growth rates) in hybrids between inbred lines of *Arabidopsis thaliana* in extreme (high temperature) environments in relation to performance in more optimal environments. Similar results were obtained by McWilliam and Griffing (1965) with maize. Another example is given by the work of Bucio Alanis, Perkins & Jinks (1969) with *Nicotiana rustica*. By comparing over a series of physical environments the magnitude of heterosis in hybrids between inbred lines, they found that heterosis decreases as the environment improves. Also, on the basis of detailed biometrical measurements they were able to predict the change in heterosis based on the sensitivity of the hybrids and that of the inbred lines to environmental variables. Parsons (1971) has pointed out that such a behaviour of hybrids may be another mechanism to store genetic variability. In effect, in areas (or at times) of little environmental stress, the fitness differential (and consequently the segregational load) between heterozygotes and homozygotes will be small and the population can expand maintaining genetic polymorphism with a relatively small genetic load; however, in areas (or at times) of environmental stress, only some especially well-adapted heterozygotes will survive. Hence the population will suffer the negative effects of heterosis (i.e. segregation of inferior genotypes) only at specific times or places (see also Hedrick, Ginevan & Ewing 1976 for a general review of this problem).

A mechanism such as that which has just been explained has direct demographic implications. Mortality due to density independent factors in ecologically marginal populations should increase over that experienced in ecologically more central populations, thereby limiting the ability of the species to expand into stressful environment. Furthermore, genetic cohesion in ecologically marginal populations will increase, thereby reducing the chances of evolving adaptations to novel ecological situations since new mutations not only have to provide immediate fitness in terms of the character or characters they affect, but have to be capable of becoming integrated immediately into the gene pool of the population. Gene flow will further enhance the integrity of the genetic structure of these marginal populations since density considerations dictate that most of the flow will be from the centre to the periphery (Antonovics 1976). These factors may explain in part why species maintain their morphological and physiological integrity in spite of a very low level of gene flow (Ehrlich & Raven 1969; Raven 1979). It also may support the contention that speciation may have to be preceeded by a 'genetic revolution' (Mayr 1963) or some other 'catastrophic' event (Lewis 1962; Lewis & Raven 1958).

But fitness is not only a function of the physical environment but of the biological environment as well. Elsewhere (Solbrig 1976) I have explained how the biological environment selects not for a uniformly superior genotype, but for superior genotypes with the ability of producing variable offspring. In effect, any genotype that is a superior competitor or has especially good defences against predators, acts as a selective agent on those competitors and predators. Consequently, with time and as a result of evolutionary changes in the community of plants and animals, every 'superior' genotype loses its competitive advantage. To avoid extinction the population must maintain the ability to generate new genetic combinations or otherwise it must be able to migrate and successfully establish in a new territory. The breeding system (to be discussed in the next chapter) and the trade-offs between seed size and number (to be taken up in Chapter 6) are important determinants of which of those two demographic strategies will be selected.

If the biological environment selects primarily for genotypes with the capacity of producing variable offspring (Levin 1975; Solbrig 1976) and the physical environment selects primarily for heterotic combinations (Parsons 1971) it is not surprising that populations of plants show a great deal of genetic variability. In effect, by and large there is a negative correlation between physical stress and biological stress. In effect, in environments with high physical stress total density tends to be low and consequently competition is less important while in environments where density and competition are high the physical stress is usually low. Both environments will favour heterozygosity but for different reasons and through different mechanisms (heterosis *v.* outbreeding). In ecologically optimal populations variability and heterozygosity *per se* would be favoured; in ecologically marginal environments heterotic and coadapted hybrid combinations would be selected. Central populations should be more variable phenotypically but not necessarily more polymorphic genetically (number of alleles/locus) than marginal populations. This hypothesis can be tested by simultaneous determination of the genetic composition of several loci (as in the barley experiment of Clegg *et al.* explained previously). In marginal populations fewer genotypes at higher frequency are expected. It can also be tested by investigating the life cycle through demographic studies as outlined in Chapter 1, which should reveal higher reproductive efforts (to account for the increased genetic load) and less density-dependent mortality in marginal than central populations.

SUMMARY AND CONCLUSIONS

It is usually assumed that plants respond to changes in the environment, especially those factors that affect the probability of survival of individuals and their offspring, by adjusting their phenotype including the life history components. The principal environmental factors to which plants have to adjust are physical factors such as light, water, minerals and temperature and biological factors, primarily predation, and competition. Although most species of plants posses the ability to respond to temporally varying conditions by adjusting their growth, development or phenological behavior, long term evolutionary adjustments involve changes in allelic frequency. The data presented in this chapter point towards the genome as a highly integrated system. Any mutation must not only have a favorable effect on a particular character but must be able to confer increased fitness in a variety of genetic backgrounds. Consequently any new mutation must satisfy a number of requirements: (1) It must code for a functional enzyme; (2) that enzyme must have appropriate physiochemical characteristics so that it can function harmoniously in the cellular environment; (3) the enzyme must affect its biochemical pathway in such a way that the resultant phenotype is functional; (4) that phenotype must have the effect of increasing the survivorship and/or reproductive potential of the individual plant in its environment, and (5) the new allele must be able to repeat steps 1 through 4 when placed by recombination in other genetical backgrounds; if not always, at least fairly frequently.

Given these factors and the stochastic character of mutation, evolutionary changes of a genetic nature will take a great deal of experimentation in the form of wasted gametes and zygotes, which has demographic implications. Furthermore energy considerations constrain the number of gametes that a plant can produce. The interplay between genetic structure and demographic parameters has not been explored in detail. It is a promising, although conceptually and experimentally difficult area.

Chapter 4
Demographic Factors and Mating Patterns in Angiosperms

DAVID G. LLOYD

In considering the progression of individuals through their life history and their turnover in a population, demography is concerned sequentially with vegetative, flower and seed biology (Chapter 1). This chapter is principally directed towards certain interrelationships between vegetative and flowering activities. Flower biology includes flower production, gamete development (or substitution) and the mating pattern. Attention will be focused on correlations between the vegetative ecology of angiosperms and their mating patterns—the parentage of zygotes. After a review of the correlations, a series of algebraic and verbal arguments will attempt to explain the reasons for the observed patterns. Seed biology is taken up in Chapter 6.

The mating pattern exhibited by the individuals of a population is determined by numerous factors. Fortunately, many aspects of the ecological and genetical relationships between mating gametes can be embodied in one parameter, the average frequency of self-fertilization in a population. This is the most convenient single measure of the mating pattern and other factors such as the distinction between autogamy (within-flower self-fertilizations) and geitonogamy (between-flower self-fertilizations), the relatedness of cross-fertilizing plants and non-randomness in cross-fertilization either contribute to the frequency of self-fertilization or can be considered concurrently with it. The importance of the frequency of self-fertilization in influencing the number and quality (average fitness) of seeds and hence the fitness of individuals in different ecological circumstances is examined below in some detail. The frequency of self-fertilization also has important effects on the genetical composition of a population (see Allard, Jain & Workman 1968; Jain 1976), but the genetical consequences of self-fertilization are largely outside the scope of this chapter. The substitution of sexual reproduction by agamospermy is considered in Chapter 5.

* I am grateful to B. Charlesworth, B. Dommée, R. Primack, M. D. Ross, G. Valdeyron and C. J. Webb for their valuable comments on drafts of the manuscript.

ECOLOGICAL AND GEOGRAPHICAL DISTRIBUTION
OF SELF- AND CROSS-FERTILIZATION

Self- and cross-fertilization are not distributed randomly among plants of different habits or habitats or among areas of the world, or even in natural populations of a single species. A number of ecological and geographical correlations have been pointed out during the last century.

DISTRIBUTION OF SINGLE SPECIES

Delpino (cited in Darwin, 1876, p. 414) suggested that dioecious plants cannot spread as easily as monoecious or hermaphrodite species, because a dioecious species cannot reproduce if a single individual reaches a new locality. But, in referring to Delpino's suggestion, Darwin 'doubted whether this is a serious evil' because hermaphrodite and monoecious species often require cross-fertilization. Henslow (1879) observed that almost all British plants that are widely distributed beyond Britain are self-fertilizing species. He attributed the wider spread of self-fertilizing species to their independence from insect pollinators. The association between self-fertilization and the distribution of species was subsequently neglected until Baker (1955, 1966, 1967) advocated that the ability of self-fertilizing individuals to start a reproducing colony was important in long-distance dispersal, particularly of immigrants to oceanic islands. He and Stebbins (1957) provided a number of examples of disjunct species in which the populations of the original areas are cross-fertilizing, while those of a secondary area are self-fertilizing. (One quoted example, *Coprosma pumila*, is incorrect, however—Lloyd & Horning 1979). It is now known that the proportion of self-incompatible species is low among immigrants to certain remote island groups—the Hawaiian (Carlquist 1966, 1974) and Galapagos Islands (Rick, 1966) and New Zealand (Carlquist 1966; Garnock-Jones 1976; Lloyd 1975b).

Among plants on continents, peripheral (*geographically* marginal) populations are more often self-fertilized than the central populations of the same species (reviewed in Grant 1975; Levin 1975; also Baker 1966; Davies & Young 1966; Ernst 1953; Rollins 1963; van der Pijl & Dodson 1966; Vasek 1968). The establishment of many peripheral populations from a single propagule may contribute to the higher frequency of self-fertilization. Peripheral populations are also often *(ecologically)* marginal populations (Antonovics 1976; H. Lewis 1973; Vasek 1968) and self-fertilization may also be fostered by lower population size (Vasek & Harding 1976) or lower density or by a paucity of pollinators under marginal conditions (Baker 1966; Davies & Young 1966).

HABIT

Darwin (1859, 1876) pointed out that in northern temperate floras and in New

Zealand the proportion of species with separate sexes is much higher among trees and shrubs than among herbs. A correlation between unisexuality and woody habits has since been recorded in angiosperm families on a world-wide scale (Sporne 1976). Increasing unisexuality with increasing plant height has been confirmed for British woodland species (Baker 1959) and also observed in Costa Rica (Bawa & Opler 1975). Among herbs, there is a further correlation between longevity and outbreeding. Henslow (1879) noted that self-fertilizing species are often annuals while their outcrossing allies are perennials. He suggested that this correlation explained A. de Candolle's observation that annual and biennial plants are generally more widely dispersed than perennials. Associations between annual habits and selfing and between perenniality and cross-fertilization have since been recorded in many groups, including grasses (Beddows 1931; Kornicke 1890, cited in Beddows 1931; Stebbins 1950, 1957), legumes (Kirchner 1905; Williams 1951), the tribe Cichorieae of the Compositae (Stebbins 1958) and the Polemoniaceae (Grant & Grant 1965). Moreover, outbreeding mechanisms are more common in polycarpic perennials than in monocarpic ones (Baker 1959) and in vegetatively reproducing perennials than in those lacking means of vegetative reproduction (Stebbins 1950).

All these observations consistently reveal increasing proportions of cross-fertilizing species as the length of time between sexual generations increases.

HABITAT

An association has often been noted between self-fertilization and open, colonizing, pioneer, temporary, unsaturated, disturbed, unstable, fluctuating, unpredictable or harsh environments. The latter terms are not all synonymous, however. Here I wish to distinguish in particular between two classes of open habitats, namely unsaturated habitats and harsh habitats. Unsaturated habitats are those that have only recently become available or have recently improved and are sparsely occupied by a population for only a few generations before filling or deteriorating. Harsh habitats, however, are unfavourable for plant growth generally and may remain open indefinitely. Grime (1977) has pointed out that plants living under the two sets of open environments characteristically have different adaptive strategies. The plants of harsh environments experience high stress and often have reduced stature and slow growth rates. Plants of unsaturated environments exhibit rapid growth and high fecundity (and marked phenotypic plasticity; Baker 1965). It should also be noted that unsaturated environments may be either fluctuating (= unstable or unpredictable) or temporary (available at all for only a limited period) or both. The occupation of temporary habitats by colonizing plants requires adaptations for long-distance dispersal as well as those for residence in a fluctuating environment. Both aspects of unsaturated habitats are further discussed in subsequent sections.

A correlation between unsaturated habitats and self-fertilization was apparently first noticed by Henslow (1879), who commented that most weeds are self-fertilizing. Similar statements about weeds have been reiterated many times (e.g. Allard 1965; Baker 1965, 1974; Mulligan & Findlay 1970). Stebbins (1950) made a broader observation that self-fertilizing plants 'live in habitats characterized by a great fluctuation in climatic conditions from season to season or in pioneer associations which are constantly changing their extent and position'. Stebbins (1957, 1958) and Grant (1958, 1975) provided further examples of correlations between self-fertilization and temporary, unstable habitats. Similar associations have been observed in pteridophytes (Lloyd 1974) and some groups of micro-organisms and animals (Baker 1955; Grant 1958; Moore 1976). Recently (Grant 1975; Levin & Kerster 1974; Levin 1975), the ecological characteristics of self- and cross-fertilizing species have been respectively identified with the r-strategies of organisms of unsaturated environments and the K-strategies of organisms of saturated environments (based on MacArthur & Wilson 1967).

Self- and cross-fertilizing species are not equally common at all latitudes. Kerner (1896) and other late nineteenth and early twentieth-century naturalists (see Kevan 1972) observed that the frequency of autogamy increases towards the north in the north temperate zone. At very high northern latitudes, cross-fertilizing species are uncommon, but some species are still dependent on insect pollinators (Hagerup 1951; Kevan 1972; Mosquin 1966; Savile 1972). At the other extreme, the trees of a variety of tropical forests have a particularly high frequency of outbreeding mechanisms (Ashton 1969; Bawa 1974; Bawa & Opler 1975; Ruiz & Arroyo, in press; Tomlinson 1974). The correlation between mating patterns and latitude has also been interpreted in terms of r- and K-selection (Grant 1975; Levin 1975). But in the extreme case of the arctic floras, a paucity of insects has also been suggested as a cause of the rarity of outcrossing species (Baker 1966; Davies & Young 1966; Hagerup 1951). The arctic climate is a harsh climate, as well as an uncertain one, and we should be aware that 'the r–K trade-off might not be the most interesting or most important trade-off in many cases' (Armstrong & Gilpin 1977).

The extent to which self-fertilization occurs in harsh environments in general is not certain at present. Besides the plants of the Arctic zone, the herbs and low shrubs in the flora of Timbuktu are nearly all self-fertilizing (Hagerup 1932). Hagerup related this to the avoidance of the intensely heated ground by insects. On the other hand, Pojar (1974) has observed that most plants in four communities in harsh environments in British Columbia have mechanisms promoting outcrossing. Perhaps environments that are harsh for plants are not uniformly unsuitable for pollinating insects.

Although the plants that establish on remote islands are mostly self-fertilizing, as Baker (1955) suggested, there is a high frequency of unisexuality in several of these island groups, notably New Zealand and Hawaii (Carlquist 1966, 1974; Thomson 1880; Wallace 1876). This apparent paradox was resolved by Baker (1967) and Gilmartin (1968), who suggested that the stable environments and perennial habits characteristic of these islands caused the selection of secondary outcrossing devices. The continued high frequency of self-fertilization in the Galapagos Islands, on the other hand (Rick 1966), may be the result of their younger age or poorer insect faunas (Carlquist 1974; Linsley 1966; Rick 1966) or to their general ecology. Porter (1976) has described the Galapagos Islands as 'a jumble of open pioneer habitats inhabited by a weedy flora'—features that are elsewhere associated with self-fertilization.

GENERAL ADVANTAGES OF SELF AND CROSS-FERTILIZATION

In a previous paper, which dealt with the effects of a number of reproductive parameters on the selected frequency of self-fertilization (Lloyd 1979), it was suggested that this frequency was due to the modulation of two sets of factors which affect the fitness of self- and cross-fertilizing individuals in one population. These demographic factors, to be considered below, modify the balance between the reproductive parameters.

Self-fertilization increases the number of offspring, or more precisely the number of gametes or genes contributed to offspring, in two ways. First, a plant contributes two gametes to each self-fertilized zygote but only one to each cross-fertilization, as Jain (1976) and Solbrig (1976) have also pointed out (cf. Fisher 1941). Cross-fertilization therefore has a disadvantage or cost parallel to the 'cost of meiosis' (Maynard Smith 1971a) which gives asexual reproduction an advantage over sexual reproduction. Secondly, self-fertilization increases the total seed set in many circumstances. It has been confirmed in natural populations of a number of genera that autogamously fertilized flowers produce a greater seed set than allogamously fertilized flowers (Bawa 1974; Ganders 1975a; Levin 1972; Ruiz & Arroyo, in press; Solbrig & Rollins 1977). The ability of flowers to self-fertilize does not necessarily ensure a full seed set, however (Bawa 1974; Free & Williams 1976; Ruiz & Arroyo, in press). Self-pollination may even reduce the total seed set if it interferes with cross-fertilization in self-incompatible plants (Bawa & Opler 1975). Whether self-fertilization increases the seed set also depends on how and when it occurs (Lloyd 1979, and below).

Balanced against the numerical advantages of self-fertilization is a widespread *qualitative superiority of the progeny arising from cross-fertilization.* Darwin (1877), Knuth (1906) and Müller (1883) observed that even species with cleistogamous flowers retain some capacity for cross-fertilization. Because crossing is less efficient and more costly than selfing, they argued that cross-fertilization must provide an advantage to the offspring, even in frequently self-fertilizing species. The superiority of hybrids was attributed to the dissimilarity of the uniting gametes by Darwin (e.g. 1876). Nowadays, hybrid vigour or heterosis is thought to derive in part from a number of general biochemical properties of heterozygotes (Johnson 1976a; Parsons 1971; Roose & Gottlieb 1976). Heterosis is a general property of all or many of the individuals resulting from cross-fertilization and it may be expressed in uniform or variable environments and in the presence of either physical or biological stresses.

Particularly since the influential writings of Darlington (1939) and Mather (e.g. 1943), outcrossing has often been considered to have a different advantage, arising from an increased generation of genetical variability. There are too many versions of this hypothesis to be considered here, but it may be noted that they have in common that the greater fitness resulting from outcrossing is expressed principally or entirely in a minority of plants. This minority is either a genetic elite of rare superior individuals or is better adapted to a new environment or provides a new combination of genes in an evolutionary race with a biological adversary. Many 'control of variability' hypotheses appear to invoke selective advantages to populations, but Levin (1975) and Solbrig (1976) have recently proposed plausible hypotheses based on individual selection. The reader is referred to those articles for different approaches from that adopted here to the ecological distribution of mating patterns.

The advantage accruing from outbreeding is considered, below, to arise from heterosis rather than from increased genetical variability. The principal reason for this is the present author's view that heterosis, being an expression of the intrinsic genetical nature of organisms, is likely to be much more widespread and relentless than the benefits of variable progeny, which are entirely dependent on environmental circumstances. But, except for the models in the section on habitat-dependent selection, which demand that the advantage applies to all progeny from cross-fertilization, the following arguments could also be applied to an outcrossing advantage arising from increased genetical variability.

MODES OF SELF-FERTILIZATION

The relative advantages of self- and cross-fertilization depend on exactly how self-fertilization takes place. Three principle modes of selfing have been distinguished (Lloyd 1979). 'Prior' self-fertilization occurs when a fraction of the ovules are spontaneously self-fertilized before any opportunities for crossing. 'Competing' self-fertilization occurs when a fraction of the ovules are self-fertilized, either autogamously or geitonogamously, by the action of the pollinating agent at the same time as cross-fertilization (or instead of it). 'Delayed' self-fertilization occurs when a fraction of the ovules are spontaneously self-fertilized after all opportunities for cross-fertilization have passed. The three modes of selfing differ in their effects on the amount of cross-fertilization and the total seed set.

To derive the general circumstances under which the numerical advantage of selfing outweighs the qualitative superiority of the progeny from crosses, the fitnesses of two phenotypes differing in the frequencies of selfing were compared in individual selection models for each mode of self-fertilization (Lloyd 1978). The models deduce the 'stationary conditions' under which neither an increase nor a decrease in self-fertilization is selectively advantageous. Here it is necessary to repeat only the nature of the stationary conditions in the basic models for each mode of selfing. The proportion of available (functional but not previously fertilized) ovules which are fertilized with the aid of an external pollinating agent is e. The relative fitness of (inbred) progeny from self-fertilization, i, is the average fitness of progeny from self-fertilization, w_s, divided by the average fitness of progeny from cross-fertilization, w_x. This model of inbreeding depression is oversimplified, since it assumes a constant value. The degree of inbreeding depression is affected by the number of consecutive generations of self-fertilization (Lloyd 1979, Haynard Smith 1978). Nevertheless, a single population value for inbreeding depression will suffice here to examine the effects of demographic factors.

Prior self-fertilization affects the number of ovules available for cross-pollination. Hence the stationary conditions for prior selfing ($i = \frac{1}{2}e$) in the absence of developmental or evolutionary interactions between reproductive parameters; (Lloyd 1979)) depend on the level of pollination. Complete self-fertilization is favoured if $i > \frac{1}{2}e$, and 100 per cent cross-fertilization is advantageous if $i < \frac{1}{2}e$. In this simple situation, the stationary condition under which neither phenotype is favoured over the other is a boundary where the selected mode of fertilization switches completely. Some more complex models involving certain interactions between parameters (Lloyd 1979) or habitat-dependent heterosis (below) predict intermediate frequencies of self- and cross-fertilization. But, in all cases, self-fertilization is more likely to be favoured with lower levels of pollination and higher relative fitness of progeny from selfing.

The stationary condition for competing self-fertilization is $i = \frac{1}{2}$ (Lloyd 1979). The level of fertilization by an external agent, e, does not influence the relative advantages of competing cross- and self-fertilization, since the frequency of selfing is considered to be independent of the level of pollination. Competing selfing is advantageous if $i > \frac{1}{2}$ and disadvantageous if $i < \frac{1}{2}$. The third mode of selfing, delayed self-fertilization, is always advantageous (except with certain strong interactions between parameters) since it simply adds to the seed set obtained from crosses (Lloyd 1979).

EFFECTS OF POPULATION SIZE AND STRUCTURE

The size and structure of populations may affect the favoured frequency of self-fertilization in a number of ways by influencing the level of pollination, e, or the relative fitness of selfed progeny, i.

EFFECTS ON THE LEVEL OF EXTERNALLY MEDIATED POLLINATION

Demographic effects on the level of externally mediated pollination have the greatest impact on selection of self-fertilization in the case of prior selfing, where the externally mediated cross-pollination follows self-pollination. The circumstances affecting the level of cross-fertilization, e, can be examined in a model separating the number of visits by an animal pollinator to each flower, v, from the probability that any available ovule is left unfertilized after any one visit, l. The probability that an ovule that was not previously self-fertilized will still be unfertilized after v visits is l^v, and the frequency of cross-fertilization of the available ovules, $e = 1 - l^v$. Alternatively, suppose that the number of visits to different flowers varies according to a Poisson distribution with mean μ (cf. Straw 1972). If a single visit effects full fertilization of the available ovules in a flower, the proportion of these ovules that is cross-fertilized is the probability that a flower will be visited at least once. That is, $e = 1 - \exp(-\mu)$. With either pollination model, self-fertilization will be favoured less frequently as the number of pollinator visits increases.

Several demographic factors may reduce the number of visits to each flower and hence increase the probability that selfing is favoured. A smaller population size may reduce the attractiveness of a population to animal pollinators (H. Lewis 1973) or decrease the efficiency of stigmas of wind-pollinated species in capturing pollen (Solbrig 1976). The pollination levels may also be reduced in wind- or animal-pollinated species if the average plant density decreases (Beattie 1976; Fedorov 1966; Free 1970; Levin & Kerster 1974; Solbrig 1976; Whitehead 1969) or if the patchiness (spatial variation in density) decreases (Beattie 1976; Frankie, Opler & Bawa 1976; Ruiz & Arroyo, in press). In animal-pollinated plants, autogamy is particularly likely to be selected in less abundant species if their pollinators are promiscuous and flowers of competing plant species are available at the same time and place (Heinrich 1975; Levin 1972; Mosquin 1971). Moreover, if inter-species pollinator movements are common, the proportion of ovules unfertilized after each visit may increase, further favouring self-fertilization (Levin & Anderson 1970). All bee species that have been studied are less constant to one plant species during a single foraging trip when there are many plant species and comparatively few individuals of each available (Free 1970).

It should be noted, however, that increasing abundance of a plant can result in *diminished* levels of externally mediated pollination if large or dense populations cause satiation of the available pollinators. Three situations may be most likely to overload the pool of pollinators. First, populations of the pollinators of crop plants may be regulated outside the area of the crop and therefore pollinators need not increase as crop size increases. Bond & Pope (1947) and Simpson (1954) found that in crops of field beans and cotton, respectively, there was generally more crossing from the activities of bees in small fields than in large ones. Free & Williams (1976) observed more seeds per plant at the edges than in the centres of large field bean populations; smaller populations did not show this discrepancy. Secondly, pollinator satiation could occur in harsh habitats with a low diversity of plants and animals. This may provide another reason, in addition to the increased efficiency of wind pollination in large and dense populations (Whitehead 1969; Faegri & van der Pijl 1971) why wind pollination is frequent in communities with few species. Thirdly, a plant population may increase in size more rapidly than that of its pollinator, as in deserts after a fall of rain (O. T. Solbrig, personal communication).

EFFECTS ON THE DEGREE OF HETEROSIS

The expression of heterosis is dependent upon the extent of dissimilarity between the uniting gametes (Cress 1966; Darwin 1876; Lerner 1954; Moll *et al.* 1965). Population size and structure may modify the advantage of cross-fertilization by altering the degree of relatedness of cross-fertilizing plants. Any factor that increases the degree of inbreeding in a population will cause cross-fertilizing plants to be more closely related and may therefore be expected to reduce the level of heterosis. Such factors include reduced or varying population size, bottleneck effects, subdivision into neighbourhoods by limited gene dispersal or patchiness, variation in the reproductive output of different individuals and self-fertilization itself (Jain 1975; Levin & Kerster 1974; Nei, Maruysma & Chakraborty 1975; Nei & Syakudo, 1958; Wright 1943, 1965).

The effects of plant density on the level of heterosis expressed in a population are not straightforward, since both the average relatedness of cross-fertilizing plants and the behaviour of pollinators may change with plant density. Levin & Kerster (1969) demonstrated that as the density of *Liatris aspersa* declines, the neighbourhood size remains relatively constant because the reduction in effective density is counterbalanced by an increase in the variance of pollen dispersal distances.

DIRECT EFFECTS ON THE FREQUENCY OF SELF-FERTILIZATION

With competing self- and cross-fertilization, demographic factors can affect the frequency of self-fertilization directly by influencing the behaviour of the pollinating agent. In animal-pollinated species, a change in the flight pattern may alter the proportions of xenogamous (between-plant) and geitonogamous (within-plant, between-flower) pollinations. Assuming that pollinators usually move to one of the nearest available flowers (Levin & Kerster 1974), one would expect that an increase in the number of flowers per plant or a decrease in plant density would generally result in more within-plant movements and hence in more geitonogamous self-fertilization where this is possible (Bateman 1956; Carpenter 1976; B. Dommée, personal communication

1975; R. Primack, personal communication 1976). The frequency of self-fertilization or of intra-plant flights has been observed to increase at lower plant densities in *Cheiranthus cheiri* (Bateman 1956), cotton (Simpson & Duncan 1956), *Amsinckia spectabilis* (Ganders 1975b), *Viola* species (Beattie 1976) and male-fertile individuals of *Thymus vulgaris* (Valdeyron, Dommée & Vernet 1977). In wind-pollinated plants too, lower densities (Bannister 1965; Levin & Kerster 1974) and higher numbers of flowers per plant may result in higher frequencies of geitonogamous self-fertilizations.

In high-diversity tropical forests, the low density of each species together with the large number of flowers often present on a tree at the same time may frequently lead to high proportions of pollinator movements between flowers of the same plant (Arroyo 1976; Baker 1959; Corner 1954; Fedorov 1966; Frankie *et al.* 1976). It was formerly thought that as a result most tropical trees would be regularly self-fertilized. But it is now known that many species have floral mechanisms and pollination systems encouraging or enforcing outcrossing in spite of the apparent difficulties (Bawa & Opler 1975; Frankie 1975; Janzen 1971a). Some cross-fertilizing species achieve only low levels of seed production (Frankie *et al.* 1976). Ruiz & Arroyo (in press) have calculated that the average 'reproductive efficacy' (natural fruit set divided by fruit set in controlled crosses) was 36 per cent among outcrossing species of a secondary Venezuelan forest and 16 per cent in a primary Costa Rican forest (latter figure from data of Bawa (1974)). They attribute the higher reproductive efficacy in the Venezuelan forest to the lower species diversity and greater ease of cross-fertilization.

EFFECTS OF UNCERTAIN POLLINATION LEVELS ON SELFING BY ANNUALS AND PERENNIALS

The level of externally mediated pollination in natural populations is likely to fluctuate from year to year. Baker (1959) and Stebbins (1950) have suggested that the fitness of annuals is more sensitive to fluctuations in seed set than is the fitness of polycarpic perennials and that consequently selfing is likely to be favoured more often in annuals. The effects of varying pollination levels may be examined in a simplified model that examines the average fitness of plants in two successive years consisting of one 'good' year and one 'bad' year (Schaffer & Gadgil 1975). The level of fertilization of available ovules is e in good years and ea in bad years, where $a < 1$.

To concentrate on the fluctuations in pollination, suppose that in an annual plant the plants present in 1 year are a fixed proportion, z, of the seeds produced the previous year and seeds are not viable for more than 1 year. Over a 2-year period, the overall fitness (total seed production) of an annual plant is dependent on the joint frequencies of successive events and hence on the product of the seed set of separate years, i.e.

$$w_2 = w_g \times w_b$$

$$= a(zne)^2, \qquad (1)$$

where w_g and w_b are the fitnesses in good and bad years and n is the number of ovules produced on a plant. The maximum possible fitness in 2 years, obtained after successive good years, is $(zne)^2$. Hence the ratio of average fitness:potential fitness is $a:1$. In the extreme case when no offspring are produced in a bad year, $w_b = 0$ and $w_2 = 0$.

Suppose that a perennial plant survives and reproduces for 2 years or longer and that the seeds produced in the first and subsequent years have equal chances of contributing to the next generation. Then the seed crops of two separate years provide alternative individuals to the next generation and the total fitness over a 2-year period is the sum of the fitnesses of the separate years, i.e.

$$w_2 = w_g + w_b$$

$$= (1 + a)zne. \qquad (2)$$

The maximum possible fitness over a 2-year period is $2zne$. The ratio of average:potential fitness is therefore $\frac{1}{2}(1+a):1$. Now $\frac{1}{2}(1+a) > a$ since $a < 1$. Hence, fluctuations in seeds set reduce the fitness less for perennials than for annuals. The effect of (prior and delayed) self-fertilization in raising a fluctuating seed production will therefore be less important in perennials than in annuals, as Baker and Stebbins suggested.

This model is undoubtedly too simple. Other factors, including fluctuating plant numbers, seed storage beyond 1 year and the death each year of a fraction of the population of a perennial plant will reduce the difference described above between annuals and perennials. But annual populations almost inevitably have a more rapid turnover than perennials. Whenever selfing increases the seed set and individual fitness increases with increasing seed production, selfing is likely to be selected more often in annual than in perennial plants.

HABITAT—DEPENDENT HETEROSIS

The level of heterosis was first assumed above to be constant and then to be dependent on the degree of relatedness of the cross-fertilizing parents. In addition, there is a considerable body of evidence that in both plants and animals the advantage of outcrossing increases as the conditions under which the offspring are raised become more severe. In his experiments on the relative advantages of self- and cross-fertilization in plants, Charles Darwin (e.g. 1876, p. 288) observed that the superiority of plants from crosses was often more pronounced in crowded conditions and less evident in uncrowded conditions. Plants produced from crosses also resisted unfavourable physical conditions

better than did plants resulting from self-fertilization (Darwin 1876, p. 289).
The dependence of heterosis on cultural conditions and the occurrence of
regularly self-fertilized plants led Hermann Müller (1883) to conclude that the
Knight–Darwin Law that 'nature abhors perpetual self-fertilization' is too
extreme. Darwin himself (1876, Introduction) appears to have had some
misgivings about the law as he stated that it was 'almost universal' and
'perhaps rather too strongly expressed'. Müller judiciously replaced the law
with the 'sufficient and demonstrable assumption' that 'cross-fertilization
results in offspring which vanquish the offspring of self-fertilization in the
struggle for existence'.

In this century, experiments on a range of plants and animals have often
shown that homozygotes and heterozygotes differ comparatively little in
fitness in benign environments, and that the advantage of heterozygotes
increases progressively as either physical or biological environments become
less favourable (Allard 1965; Bailey, Rees & Jones 1976; Dobzhansky 1959;
Lerner 1954; Levin 1970; Parsons 1971; Rehfeldt & Lester 1969).

Here we examine models in which the level of heterosis depends on the
nature of the environment encountered by plants. Two kinds of environment
are considered, at first separately and then in a mixed habitat.

UNSHARED SITES

A plant grows in an 'unshared site' if, during its lifetime, no other plants occur
in the area that it utilizes when fully mature. If all plants of a population grow
in unshared sites, there is no competition between them and the advantage of
outcrossing is assumed to be density-independent. Individuals growing in
unshared sites are assumed to experience a number of independent events,
during each of which the average fitnesses of plants resulting from cross- and
self-fertilizations are, respectively, o and $i_1 o$, where $i_1 \leqslant 1$. If all plants are
subjected to the same number, k, of selection events, the total fitnesses in
unshared sites of plants from cross- and self-fertilizations, w_{xu} and w_{su}, are the
products of their fitnesses in single events. That is,

$$w_{xu} = o^k, \text{ and } w_{su} = (i_1 o)^k = i_1{}^k w_{xu}.$$

The relative fitness of progeny from self-fertilization after k events,

$$i_k = \frac{w_{su}}{w_{xu}} = i_1{}^k. \tag{3}$$

As k increases, the logarithm of i_k decreases proportionately. Increasing
physical stress therefore decreases the relative fitness of offspring from self-
fertilization and increases the advantage of outcrossing.

The fitnesses of plants from cross- and self-fertilization in single events, o
and $i_1 o$, may represent either their viabilities in a suboptimal environment or
relative growth rates proportional to their eventual fecundities. Provided their

effects are independent, the separate selection events may occur simultaneously or in succession. The selection events may also be considered as increasing *levels* of stress induced by one environmental factor at one time if higher levels of stress cause an increasing number, k, of separate effects on physiological and developmental processes.

This model could also be applied to the effects of competition between species on self-fertilization, if i_1 represents the relative fitness ($\bar{w}_s/\bar{w}_x$) of progeny from self-fertilization in competition with an individual of another species, and k is the number of individuals of other species that each individual competes with during maturation. Then increasing interspecific competition, like increasing physical stress, will increase the degree of heterosis and the advantage of outcrossing.

SHARED SITES

Shared sites are defined as areas which receive a number of conspecific seeds but are able to maintain only one adult plant. The model examined here proposes that two of the y plants growing in a shared site compete initially and one of the plants dies. After this first encounter, another pair of plants from the survivors compete and another plant dies, and so on until one victor remains and competition ceases. Of the y competitors in a shared site, a fraction s is derived from self-fertilization and a fraction $1 - s$ from cross-fertilization. All plants from self-fertilization are assumed to be identical, as are all those from cross-fertilization. Hence the outcome of an encounter between two plants resulting from self-fertilization (or between two plants from crosses) is simply that one of the pair survives. In a single encounter between a plant derived from self-fertilization and one from cross-fertilization, the relative probability that the 'selfed' plant will survive compared with probability that the 'crossed' plant will survive is i^*, which is independent of the number or nature of previous encounters. We assume that selfed plants win between none and half their encounters with crossed plants and therefore $0 \leqslant i^* \leqslant 1$. What is the probability that any single selfed seed will eventually win the entire competition at a shared site compared with the probability that any single crossed seed will win? That is, what is the value of $i_y = w_{sa}/w_{xa}$, where w_{sa} and w_{xa} are the average fitnesses of selfed and crossed seeds arriving at shared sites and i_y is the relative fitness of selfed progeny after $y - 1$ encounters among y plants?

A general solution to this question was obtained by Dr Murray Smith. His procedure and general solutions for the extreme cases (initially one selfed plant and the remainder crossed plants in a competition, and initially one crossed plant and the remainder selfed plants) are given in the Appendix. The results graphed in Figure 4.1 are the geometric means, $\bar{i}_y$, of these two extremes.

When a selfed plant never wins an encounter with a crossed plant ($i^* = 0$),

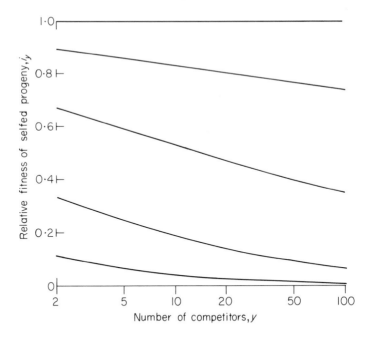

FIGURE 4.1. Graph of the average relative fitness of progeny from self-fertilization ('selfed progeny'), $\bar{I}_y$, against the number of competitors at a shared site, y, (on a logarithmic scale). The relative fitness of selfed progeny is the probability that a single 'selfed' plant will succeed in eventually occupying the shared site by itself compared with the probability that a single 'crossed' plant will do so ($\bar{I}_y = \bar{w}_s/\bar{w}_y$) after $y-1$ successive contests between y competitors. The value of $\bar{I}_y$ depends principally on the number of competitors and the relative probability of survival of a selfed plant in a single contest with a crossed plant, i^*. The curves for $i^* = 0, \frac{1}{9}, \frac{1}{3}, \frac{2}{3}, \frac{8}{9}$, and 1 join calculated values for $\bar{I}_y$ when $y = 2, 3, 4 \ldots 100$. The calculated $\bar{I}_y$ values are the geometric means of exact solutions for the two extreme possibilities for any i and y values, when the numbers of selfed plants in a competition are 1 and y $-$ 1 respectively. See text for further explanation.

then obviously selfed plants never win the entire competition for a shared site if there is one or more seeds from crossed plants on the site, i.e. $\bar{I}_y = 0$ if $s < 1$. If selfed plants win half their single encounters with crossed plants, then individual selfed and crossed plants have equal chances of being the eventual victor ($\bar{I}_y = 1$ when $i^* = 1$, regardless of the initial proportions of selfed and crossed plants and the numbers of competitors at a site). When $0 < i^* < 1$, the results are more complex. If $y > 2$, then $\bar{I}_y < i^*$, and as y increases the relative fitness of selfed progeny progressively decreases to $i_\infty = 0$. And, for any particular number of competitors at a site, $\bar{I}_y$ decreases as i^* decreases.

Figure 4.1 shows the $\bar{I}_y$ curves for $i^* = 1, \frac{8}{9}, \frac{2}{3}, \frac{1}{3}, \frac{1}{9}$ and 0 (the probability of a selfed plant winning one encounter with a crossed plant is, respectively, 50, 47, 40, 25, 10 and 0 per cent). It is evident that if the number of competitors is high, selfed plants have low probabilities of winning an entire competition at shared sites, unless their competitive ability in single encounters is close to that of

variety of reproductive features and effects of hybridization and competition from other gene pools are significant (Antonovics 1968, Jain 1976; Levin 1972; Lloyd 1979 and unpublished). For ease of presentation, the various demographic factors have been treated separately. But, in nature, diverse selective forces are likely to be simultaneously influencing the mating pattern of a population. The various factors influencing selection of mating patterns can be viewed as effecting a compromise in individual selection between advantages of self-fertilization in increasing the number of gametes contributed to zygotes and an advantage of cross-fertilization in increasing the average fitness of zygotes.

The post-fertilization advantage of crossing is attributed herein to heterosis. The magnitude of heterosis is itself influenced by the genetical structure of populations and the proportions of plants which are eliminated in circumstances allowing selection between the products of self- and cross-fertilization. In the models presented above, heterosis in both shared and unshared sites is postulated to be less in benign environments and to become greater as genotype-dependent elimination becomes more severe. This conclusion is parallel to views expressed nearly a century ago by H. Müller (1883) and others and is in accord with the experimental evidence reviewed (see also Chapter 3).

Another view has predominated during the last 40 years, however. Grant (1958, 1975), Mather (1940 and later), Stebbins (e.g. 1950, 1970) and others have maintained that in unsaturated environments the average fitness of the progeny from self-fertilization is higher than that of progeny from crossings. Stebbins (1970) labelled as the 'infective principle' the view that the increase of a colonizing plant 'will take place most effectively if its descendants share all of the genetic qualities that have adapted it particularly well to that specific habitat'. In my opinion, there is neither evidence nor theoretical justification for supposing that the progeny from selfing are better adapted than those from crossing in these situations.

The advantages of self-fertilization are better attributed to increased success in fertilization. Because two gametes are contributed by a plant during self-fertilization and only one is contributed to a cross-fertilized zygote, self-fertilization has an invariable advantage in individual selection, but not in group selection (Lloyd 1979). Another, and ecologically more significant, advantage of selfing stems from the increased seed set it provides in many circumstances. It is useful to recognize two situations (extremes really) where an increased seed set is advantageous. First, an 'autogamy of defense' (Bocquet 1968) may be advantageous when the average fitness of cross-fertilized seeds is much higher than that of self-fertilized seeds, but cross-pollination is insufficient or unreliable. Hermann Müller expressed this advantage in 1883 when he wrote that 'though cross-fertilization is better than self-fertilization, yet self-fertilization is infinitely better than the absence of fertilization and consequent sterility'. Secondly, an 'autogamy of conquest' (Bocquet 1968) is advantageous in unsaturated environments when the average fitness of cross- and self-fertilized zygotes is more similar and the number of propagules produced is particularly important.

The frequency of self-fertilization is not selected in isolation from the selection of other characters. A number of adaptive mechanisms, such as those relating to pollination and seed dispersal syndromes, and adaptive strategies relating to reproductive effort, phenology, etc. may incidentally influence the frequency of self-fertilization in a population and prevent the 'optimal' fre- quency from being freely selected. The direct effects of plant size and density on pollinator behaviour mentioned earlier cause the frequency of self-pollina- tion to vary regardless of whether this is selectively advantageous. In many plants, self-fertilization by geitonogamy may not be preventable (Arroyo 1976; Frankie *et al.* 1976). A striking instance where most pollinator move- ments lead to unprofitable incompatible pollinations has been described by Frankie *et al.* (1976).

The models presented above predict a variety of correlations between the mating pattern of a population and various ecological characteristics. But to what degree do the factors described here actually cause the ecological and geographical patterns observed? It is apparent, as Jain (1976) pointed out, that there is an 'embarrassment of riches' to explain the correlations. Frequent self-fertilization has been predicted for separate reasons under conditions of pollinator paucity (Lloyd 1979), in small, scattered and inbred populations, in annuals, in unsaturated (fluctuating) habitats and in temporary habitats. Some of these circumstances are themselves complex. Further factors not considered here have been postulated, notably in several distinct versions of hypotheses based on the control of genetical variability by breeding patterns (e.g. Grant 1975; Levin 1975; Solbrig 1976). There is now a need to unravel the relative importance of the various forces postulated.

Up to the present time, discussions of the ecological and geographical distribution of mating patterns have usually searched among the observed correlations for evidence for the operation of one or another selective force. This approach has opened the subject up, but it has severe limitations. Correlations can never provide compelling evidence for causal relationships, and their value here is diminished by the numerous correlations between the pertinent ecological factors themselves. Pollinator deficiencies, reduced popu- lations, annual habits and unsaturated and temporary habitats may all on occasions be associated with each other.

Detailed observations of the actual selective forces involved will be required to determine the importance of the various factors contributing to the multiple correlations between mating patterns and ecological parameters. Much more data are needed on modes of self-fertilization, levels and patterns of pollinator activity, the numbers and fates of seeds produced, the propor- tions of seedlings which mature to adulthood (as a measure of opportunities for selection), levels of heterosis and population structure and persistence. At present the data available are woefully insufficient to discriminate between the hypotheses relating mating patterns and demography.

APPENDIX

M. H. SMITH,

Department of Mathematics, University of Canterbury, Christchurch, New Zealand

The model used for calculating the relative fitness of seeds from self-fertilization assumes that the seeds at a shared site compete sequentially in random pairs until one seed, the 'victor', remains. We shall denote by $p_y(x)$, the probability that the victor is a 'selfed' seed when initially there were y seeds sharing a site, of which x were from self-fertilization. Let $q = i^*/(i^* + 1)$ denote the probability that a selfed seed overcomes a 'crossed' seed in any single encounter. The equations for $p_y(x)$ are then

$$y(y-1)p_y(x) = [x(x-1) + 2x(y-x)(1-q)]p_{y-1}(x-1)$$

$$+ [(y-x)(y-x-1) + 2x(y-x)q]p_{y-1}(x), \quad \text{(A1)}$$

and

$$p_y(0) = 0, \quad p_y(y) = 1, \quad \text{(A2)}$$

for $1 \leqslant x \leqslant y+1$.

Since each individual selfed seed has the same probability of being the victor, this probability will be $p_y(x)/x$. Thus, the relative fitness of an individual selfed seed compared with that of an individual crossed seed when there are initially x selfed seeds among a total of y seeds is

$$i_y(x) = \frac{(y-x)p_y(x)}{x(1-p_y(x))}.$$

While the numerical solution of (A1) subject to boundary conditions (A2) is simple on a computer, it is not possible to get an analytic solution for general y and x. However, some useful analysis of the relative fitnesses can be made.

From the symmetry of the situation with regard to selfed and crossed seeds, it follows that

$$i_y(y-x) = 1/i'_y(x), \quad \text{(A3)}$$

where $i'_y(x)$ denotes the fitness of an individual selfed seed relative to an individual crossed seed but where q has been replaced by $1-q$ (or equivalently, where i^* has been replaced by $1/i^*$).

It can easily be verified from (A1) that

$$p_y(1) = \prod_{k=2}^{y} \frac{k-2+2q}{k},$$

and hence it follows that

$$i_y(1) = (y-1) \Bigg/ \left[\prod_{k=2}^{y} \frac{k}{k-2+2g} - 1 \right], \tag{A4}$$

and then using (A3)

$$i_y(y-1) = \frac{1}{(y-1)} \cdot \left[\prod_{k=2}^{y} \frac{k}{k-2q} - 1 \right]. \tag{A5}$$

Intuition suggests that $i_y(x)$ should increase in x, and this can be proved by induction. Consequently (A4) and (A5) give the extreme values for $i_y(x)$. The geometric mean of $i_y(1)$ and $i_y(y-1)$ has been chosen because it will be close to the middle value of $i_y(x)$ ($i_y(y/2)$ for y even). This is because it can be shown that $i_y(x)$ is a convex function of x for $i^* \leqslant 1$(and concave for $i^* \geqslant 1$). Thus we define

$$\bar{i}_y = \sqrt{i_y(1)i_y(y-1)}.$$

A simple approximation for $\bar{i}_y$ when y is large now follows from the results

$$i_y(1) \sim \frac{1}{\Gamma(2q)} y^{2q-1}$$

and

$$i_y(y-1) \sim \Gamma(2-2q) \, y^{2q-1}.$$

Thus

$$\bar{i}_y \sim \sqrt{\frac{\Gamma(2-2q)}{\Gamma(2q)}} \, y^{2q-1},$$

which on making an approximation for the expression in the square root, gives, for $q \leqslant \frac{1}{2}$ ($i^* \leqslant 1$),

$$\bar{i}_y \backsimeq \frac{1}{\sqrt{2-2q}} = y^{2q-1}.$$

Chapter 5
Demography and Vegetative Reproduction

WARREN G. ABRAHAMSON

The objective of this chapter is to consider vegetative reproduction from an evolutionary and ecological viewpoint. No attempt is made to review all the existing literature on vegetative reproduction since this task is being dealt with elsewhere (Werner in prep.). My aim here is to discuss the special problems that vegetative reproduction presents for both the theoretical and the field and experimental ecologist.

Asexual reproduction refers to any means of propagation that does not involve genetic recombination. Since asexual reproduction by-passes pollination, there is no dependence on other organisms, making vegetative reproduction a very efficient mode of propagation. On the other hand, because no genetic recombination is involved, asexual reproduction leads to the production of offspring that are genetically identical to the parent. In this respect, vegetative reproduction is more akin to growth than to reproduction. Consequently, in any treatment of asexual reproduction the question arises as to whether it should be considered as a form of reproduction, that is, a mechanism of producing descendants leading to a new generation in the demographic sense. When a strawberry plant is growing it produces runners and runner plants. It can be argued that the runner plants are no more offspring than are worker ants or the new branches produced by a tree in spring. But these runner plants are capable of independent life and, upon the death of the parent plant, propagate the maternal genes in time and space, which is the function of reproduction.

The confusion arises from attempting to use a model of reproduction developed from observing and studying animals, especially higher vertebrates. If we represent by N the population of individuals in a human population or that of any higher vertebrate, it can be easily seen that it is formed by a number of individuals n_1, n_2, n_3, ... n_n, each of them genetically distinct. Such a population will give rise in time to a new population N', formed by individuals n_1', n_2', n_3', ... n_n', again genetically distinct from each other and their ancestors, carrying the genes of their ancestors but not their genotypes. In most plants the same situation can be identified. In a population N an array of distinct genotypes can be identified that will give rise by sexual reproduction to a genetically distinct offspring population N'. In addition in many species (potentially all) genetically distinct individuals can also give rise to genetically identical (to each other and their ancestor), but physiologically independent offspring. If we represent these genetically identical individuals by the letter V, then the population of plants N is formed by a number of genetically distinct clones n_1, n_2, n_3, ... and each of them in turn is represented by a set $n_1 = v_1 + v_2 + v_3 \ldots v_v$ of physiologically independent, but genetically identical, individuals. Consequently a population of vegetatively and sexually reproducing plants is represented by

$$N = n_1 + n_2 + n_3 + \ldots + n_n$$

$$N' = n_1' + n_2' + \ldots n_n' = (v_1 + v_2 + v_3) + (v_4 + v_5) + (v_6 + \ldots v_v) + \ldots$$

Vegetative reproduction is therefore neither equivalent to sexual reproduction (that usually continues to take place) nor growth of parts such as a new branch, but a distinct and well-defined phenomenon. The closest analogy in the animal kingdom is a colony of ants or bees. However, since worker bees are unable to reproduce themselves the analogy is not entirely correct. Following Harper (1977) and Harper & White (1974) we will call the genetically distinct individuals, n, in a population *genets*, while the genetically identical individuals, v, arising by vegetative reproduction of a genet will be called *ramets*.

KINDS OF ASEXUAL REPRODUCTION

Two principal kinds of asexual reproduction are generally recognized:

(1) *Agamospermy*, which includes all types of asexual reproduction which tend to replace or act as substitutes for sexual reproduction by seed (Stebbins 1950). The term *apomixis* is often used instead of agamospermy, but strictly speaking apomixis is synonymous with asexual reproduction. Agamospermy can be either facultative, obligate or pseudogamous (no fertilization takes place but pollination is still required for seed development).

(2) *Vegetative reproduction*, which is reproduction by structures other than seeds (such as runners, bulbils, rhizomes, etc.). In this chapter I will deal primarily with vegetative reproduction rather than agamospermy, but the general adaptive significance of asexual reproduction is essentially the same regardless of the method.

The methods by which plants vegetatively propagate are fairly numerous. The following list is perhaps not complete but will point out the diversity of morphological structures involved.

(a)   Stolons and runners. These are long slender stems that grow along the surface of the soil sending down adventitious roots and producing new shoots, as in strawberry (*Fragaria* spp.) or in crab grass (*Dactylis glomerata*).

(b)   Rhizomes—essentially similar to stolons, only found underground with much reduced leaves (underground stem), for example *Iris,* some ferns, Kentucky bluegrass (*Poa pratensis*) and golden rods (*Solidago* spp.).

(c)   Tubers—strongly thickened rhizomes with little vascular tissue, mostly storage parenchyma and axillary buds, for example 'Irish' potato (*Solanum tuberosum*), sweet potato (*Ipomoea batatas*) and yam (*Dioscorea* sp.).

(d)   Bulbs—large underground buds made up of a stem and modified thickened leaves, for example tulip (*Tulipa* spp.), onion (*Allium cepa*), daffodil (*Narcissus pseudo-narcissus*) and lily (*Lilium* spp.) (also secondary bulbils or bulblets which can be produced in inflorescences, leaf axils and underground).

(e)   Corms—enlarged underground stems covered with one or more layers of leaf bases (differs from bulb in that food storage is in the stem, not the modified leaves), for example *Gladiolus, Crocus* and *Cyclamen.*

(f)   Roots and stems—some species produce 'suckers' or sprouts, for example cherry, apple, raspberry, blackberry and dandelion; others produce slender, long, horizontal roots which send up shoots, for example Canada thistle, black locust and silver poplar; while others develop adventitious roots near tips of branches with new shoots arising from axillary buds, for example blackberries and dewberries (*Rubus* spp.), spruces (*Picea* spp.), hemlock (*Tsuga canadensis*) and arbor vitae (*Thuja occidentalis*).

(g)   Leaves and adventitious buds—plantlets arise from meristematic tissues on leaf margins (e.g. *Kalanchoe*), on leaf tips (e.g. walking fern (*Camptosorus rhizophyllus*) or in inflorescences (field garlic, *Allium vineale*).

(h)   Fragmentation—spread and dispersal of various vegetative plant parts (e.g. black willow, *Salix nigra*).

(i)   Cauline rosettes—rare, involves the formation of a rosette of leaves on the inflorescence stalk after seed formation (for example *Barbarea vulgaris,* see MacDonald & Cavers, 1974).

OCCURRENCE OF VEGETATIVE REPRODUCTION

The occurrence of vegetative reproduction is common in lower plants (e.g. fission, budding, gemmae, bulbils, deciduous leaves or branches, spore formation, etc.). However, this discussion will be restricted to the occurrence of vegetative propagation in higher plants and primarily herbaceous higher plants.

The frequency of vegetative reproduction in flora is variable. Söyrinki (1938) examined the angiosperm flora of the alpine region of Petsamo, Lapland and found that no fewer than 45 per cent of all species reproduced primarily by vegetative means. Perttula (1941) investigated the anigosperm vegetation of Bromarv parish on the northern side of the Finnish Bay (typical forest and meadow of southern Finland) and found 80 per cent of the species possessed the means for vegetative reproduction. Salisbury (1942) pointed out that two-thirds of the perennial species common in most counties and vice-counties of Great Britain possess pronounced vegetative reproduction.

Much has been made of the common occurrence of vegetative reproduction at high latitudes or high altitudes. Mooney & Billings (1961) showed for *Oxyria digyna* a north to south cline of increased rhizome production. They argued that rhizome production may have survival value in the unstable substrates (disturbed habitats resulting from soil movement) of the northern habitats. There are also examples like *Polygonum viviparum* growing on Mount Everest at 4 400 m (14 500 ft) which reproduces by bulbils (Salisbury 1942). *Rubus chamaemorus* multiplies chiefly by means of suckers in exposed southern situations as well as in the arctic (Resvoll 1925, 1929). Bulbiferous plants (e.g. *Saxifraga cernua*) are common in arctic regions (Salisbury 1942). It has been argued (e.g. Salisbury 1942) that seed reproduction typically requires higher temperatures than those necessary for vegetative growth. Additionally, pollinators may be fewer in these extreme habitats, decreasing the probability of seed set.

In the tropics sexual reproduction is more common than asexual except in some cases of extreme environments such as the elfin forest of Puerto Rico (with environmental characteristics similar to those in plant propagator mist tents) (Nevling 1971). For example, *Tabebuia rigida* is probably represented in the elfin forest located on Pico del Oeste of Puerto Rico by only one or a few clones. The effect of the peculiar environment of the elfin forest has been to permit or perhaps even promote the amount of vegetative reproduction (Gill 1969; Nevling 1971). Vegetative reproduction habitually occurs in lianas of the primary tropical forest of Veracruz, Mexico, but not in the trees of the same forest. Among these vines there is a range of vegetative propagation methods, with varying amounts of physiological dependency, from the most dependent shoot suckers to the most independent orthotrophic shoots arising near the ends of long stolons or rhizomes (Peñalosa 1975).

The dependence of boreal forest herbs on vegetative reproduction has also been noted. Vegetative reproduction is more important in the maintenance of local populations of woodland herb species than seed reproduction (Anderson 1965; Anderson & Loucks 1973). Noteworthy examples include *Dentaria bulbifera, Nasturtium sylvestre, Ficaria verna* var. *bulbifera, Mercurialis perennis, Trientalis borealis, Maianthemum* spp., *Uvularia perfoliata*, various Orchidaceae, etc. and *Primula* spp. (Anderson & Loucks 1973; Kawano, Ihara & Suzuki 1968; Kawano, Suzuki & Kojima 1971; Kawano 1975; Salisbury 1942; Struick 1965; Tamm 1972a, b; Whigham 1974).

It is also typical to encounter species with well-developed vegetative propagation in aquatic habitats. *Acorus calamus*, for example, was introduced to Europe by man where it is sterile, due to triploidy. Torn-off rhizomes function to disperse these species to new sites (van der Pijl 1972). In other aquatic plants, fragmentation of stems and stolons can often take over the propagation and dispersal function of seeds. Examples include *Elodea canadensis* in Europe (only one sex was introduced or maintained), *Lemna* spp. (where flowering is rare, dispersal is effected often by the feet of waterfowl), *Pistia stratiotes* in the tropics and *Eichhornia crassipes* (an all-too-well-known example which is sterile in introduced areas) van der Pijl 1972). Some water plants produce hardy, dormant buds (gemmae, hibernacula, winter buds or turions) which sink to the mud bottom to overwinter (e.g. *Utricularia, Myriophyllum, Potamogeton, Cymodocea antarctica, Nymphaea micrantha*) (van der Pijl 1972). Still others increase spectacularly by vegetative means perhaps due to the reduced resistance of mud for penetration and the water transportation of the bulky propagules or fragments (e.g. *Carex paludosa, Hippuris vulgaris, Typha* spp., *Stratiotes aloides, Littorella lacustris, Hydrocharis morsus-ranae* and *Cladium jamaicense*) (McNaughton 1975; Salisbury 1942; Steward & Ornes 1975). Perhaps the most notable example is the success of *Spartina anglica* since 1890 (an allopolyploid, $2N = 122$), which was naturally derived from the sterile hybrid *S. × townsendii* ($2N = 62$) (Huskins 1931).

Ecosystems which are influenced to an appreciable degree by fire include numerous plant species that are strong vegetative reproducers. Some species show spectacular increase due to fire-induced vegetative propagation. Table 5.1 illustrates two such examples. In both species the rapid population build-up is due to sprouting from rhizomes (present before the burn), not seeds. Excavation of underground parts shows these rhizomes to be extensive and large relative to above-ground shoots. Thus, these fire-adapted species exhibit what might be called a 'sit and wait' strategy, in that they apparently survive with little above-ground biomass for long periods of time before fire causes rapid nutrient release and/or release from shading (Abrahamson, personal observation). This strategy might allow for rapid uptake of nutrients (well-developed below ground organs) after a fire to take advantage of increased light levels (somewhat analogous to so-called annual 'phoenix' plants of fire-adapted habitats).

On the other hand, fires may frequently stimulate vegetative reproduction of the prevailing dominants to such an extent that they physically compete with and ultimately eliminate species with low densities and dominance. Dense monotype grassland or marshland (dominated by Cyperaceae, Typhaceae, Juncaceae) may form due to burning (Vogl 1974). Post-fire regeneration via vegetative means appears widespread in numerous ecosystems including Mediterranean climate regions (Hanes 1971; LeHoverou 1974; Naveh 1974), sagebrush-grass range and semi-desert grassland (where rhizomatous species—grasses, forbs and sprouting shrubs—come back faster than those that must re-establish by seed; Blaisdell 1953; Humphrey 1974; Jepson 1939), temperate deciduous forest (e.g. *Betula papyrifera, Populus tremuloides* in northern United States; Ahlgren 1974; Horton & Hopkins 1965; Lutz 1956; Rowe 1955), grasslands (Vogl 1969, 1974) and northern spruce forest (e.g.

TABLE 5.1. Summary of vegetational measurements for two species of a burned grassy swale area in south-central Florida (Archbold Biological Station, Lake Placid, Florida). Data obtained using a modified line intercept method 200 metres in length. Burn occurred January 21, 1977 (Abrahamson, unpublished data).

Species	Measurement time	Shoot density/ hectare	Dominance (% cover)	Weighted* frequency
Gaylussacia dumosa	9 days Preburn	0	0	0
	103 days Post-burn	86,278	1·02	9·69
	179 days Post-burn	91,274	1·42	7·34
	360 days Post-burn	141,555	1·46	11·32
	726 days Post-burn	11,389	0·13	1·52
Vaccinium myrsinites	9 days Preburn	2,444	0·08	0·37
	103 days Post-burn	32,124	0·58	4·97
	179 days Post-burn	67,465	1·91	6·91
	360 days Post-burn	52,950	1·81	6·10
	726 days Post-burn	27,739	1·38	4·02

* Weighted according to size of shoots

Vaccinium spp., *Ledum groenlandicum, Kalmia* spp., *Cornus canadensis, Alnus rugosa, Clintonia borealis, Equisetum sylvaticum, Maianthemum canadense;* Beisleigh & Yarranton 1974).

Grazed ecosystems are composed of numerous species with vegetative means of propagation. Species in these habitats have a reduced opportunity for seed reproduction and consequently vegetatively propagating perennials are favoured. Examples include various members of the Cactaceae (Glendening 1952; Schaffner 1938) and numerous grasses and forbs (Abrahamson 1975a; Branson 1955; Harris & Brougham 1968). The endemic Koa tree (*Acacia koa*) of the Hawaiian islands and its interaction with introduced goats is an example of the importance of the pre-adaptation of vegetative propagation to survival under grazing pressure (Spatz and Mueller-Dombois 1973).

Many weedy as well as many widespread species have been shown to rely heavily on vegetative reproduction. Some of the more notorious examples include *Equisetum* spp. (Salisbury 1961), many ferns such as *Gleichenia* spp. in the tropics (Holttum 1938) or the bracken, *Pteridium aquilinum* (Braid 1948), *Tridax procumbens* (Adams & Baker 1962; Baker 1965, 1974), *Pennisetum clandestinum* (Sanchez & Davis 1969), and *Rubus fruticosus* in New Zealand or *Opuntia inermis* in Australia (Salisbury 1942). Salisbury (1942) showed that many widespread species often have vegetative multiplication as an alternative and argued that these species have a considerable range of tolerance of climatic conditions and efficient reproduction under suboptimal conditions. Clausen, Keck & Hiesey (1947) suggested that plants with efficient vegetative reproduction are better equipped to tolerate extreme environmental conditions.

Finally, sterile hybrids and polyploids are often found to reproduce through vegetative means (agamospermy as well as vegetative reproduction). *Equisetum* × *ferrissi* is one of the most abundant and vigorous horsetails in North America and yet is a completely sterile hybrid. The allopolyploid *Spartina anglica* was discussed above. Correlation is seen between sterility and the presence of vegetative propagation in *Allium, Solanum tuberosum* and *Lilium bulbiferum* (van der Pijl 1972), and in single, self-incompatible clones in *Acorus calamus, Butomus umbellatus, Lysimachia nummularia* and *Veronica filiformis* in Europe (Faegri & van der Pijl 1971). Clausen, Keck & Hiesey (1947) and Stebbins (1950) argue that vegetative reproduction allows maintenance of populations of perennial polyploids (both auto- and allo-) in spite of their typically inferior seed. Genera showing this relation include *Antennaria, Arnica, Artemisia, Biscutella, Fragaria, Hieracium* and *Rubus* (Stebbins 1950).

ADVANTAGES OF VEGETATIVE REPRODUCTION

Bonner (1958), Maynard Smith (1971b) and Williams (1975) in their discussions of sexual and asexual reproduction pose the question, 'What use is sex?' Williams (1975) states that sex is a parental adaptation to the likelihood of offspring having to face changed or uncertain conditions. While comparative evidence supports this conclusion as to what sex is adaptive to, it does not show, in a cost-benefit way, why it is adaptive. Williams (1975) points out the resulting paradox. In terms of effectiveness of gene transmission, sexual reproduction appears to be at a distinct and enormous disadvantage. An asexual parent has double the genetic representation of a sexual parent in the offspring generation. However, many organisms utilize both means of reproduction at evolutionary equilibrium indicating that the average return per investment is the same for both processes.

If it were more expensive to produce progeny by one mode, the species would lower its investment in that mode through evolution until the difference was eliminated (see Competition and Density section). Williams (1975) argues that this difference would be eliminated if the cost-benefit relation were frequency dependent (less efficient becomes more efficient as it becomes less frequent).

The importance of vegetative reproduction to the maintenance of local populations is well documented (e.g. Anderson & Loucks 1973; Bradbury & Hofstra 1976; Steward & Ornes 1975; Whigham, 1974). Where both vegetative and sexual reproduction occur simultaneously, the vegetative offspring will develop immediately, and quickly become an adult, usually with a larger food supply than that associated with seed formation. This larger supply may be due to (a) the larger size of the detached propagule (i.e. bulbils of *Dentaria bulbifera*) or (b) the prolonged attachment of the vegetative offshoot (i.e. root suckers of *Rubus*). Sexual reproduction frequently provides for dormant and widely dispersed propagules. Williams (1975) summarized the expected contrasts between asexual and sexual reproduction (Table 5.2).

Vegetative reproduction in flowering plants can also be considered as a low-risk mechanism to proliferate the genet by producing physiologically independent ramets. A striking example is seen in aspen trees which sucker from roots forming clones of over 40 hectares and perhaps attaining 8 000 years in age (Cottam 1954). Recruitment by vegetative means is independent of external pollination so that single plant is completely equipped to colonize a given area. Ramet populations can expand or contract depending on the current environmental conditions (Sarukhán, 1976).

Another advantage of vegetative reproduction is the increased longevity and perenniation of the genet. Ramet formation is not ancillary to sexual reproduction but is a system allowing for the maintenance of certain gene

TABLE 5.2. Expected differences between asexually and sexually produced offspring (after Williams 1975).

Asexual offspring	Sexual offspring
Mitotically standardized	Meiotically diversified
Produced continuously*	Seasonally limited
Develop close to parent†	Widely dispersed
Develop immediately	Dormant
Develop more directly to reproductive stage‡	Develop more slowly through a non-reproductive stage
Environment and optimum genotype predictable from those of parent	Environment and optimum genotype unpredicatable
Low mortality rate	High mortality rate

* Some plants produce vegetative offshoots seasonally (*Rubus, Fragaria, Solidago*) while others produce continuously (*Lemna, Spirodela*).
† Many plant vegetative offshoots have the potential for dispersal (e.g. *Allium* bulbils, *Acorus* rhizomes, *Eichhornia* or *Lemna* plants) (see van der Pijl 1972).
‡ Vegetative offspring show much higher growth rate than seedlings (e.g. Whigman 1974).

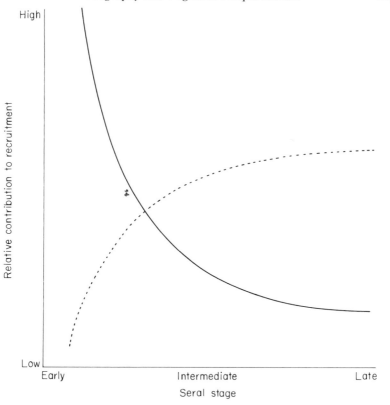

FIGURE 5.2. The relative contribution to recruitment of seed reproduction (———) and vegetative reproduction (– – – –) in herbaceous plants. At the extreme left, the environment is unpredictable favouring seed reproduction, while at the extreme right, the environment is competitive favouring reproduction for local recruitment.

PROBLEMS OF APPLICATION OF DEMOGRAPHY TO VEGETATIVE REPRODUCTION

Harper (1967) pointed out that two interlinked properties of higher plants have seriously hindered the development of plant demography:
1 plasticity and
2 vegetative reproduction.

He cites the plasticity of seed production in, for example, *Chenopodium album* (4 seeds or 100 000) and the problems which develop where ramets remain connected to the parent. When is the ramet counted as an individual?

Vegetative reproduction confounds the defining of a life span for a plant when attempting to relate the optimal age of first reproduction to mean longevity (Schaffer & Gadgil, 1975). It is difficult if not impossible to know the age of vegetatively propagating plants. Additionally Werner (1975b) has shown that the size of plants (in *Dipsacus fullonum*) and not age is more reliable for making predictive statements concerning the death, survival, or reproduction of individuals.

The life expectancy of plants originating from seed is frequently less than that of vegetatively produced plants (Langer 1956; Sarukhán & Harper 1973; Thomas & Dale 1974). This has the consequences of modifying the survivorship curve. Species dependent on sexual reproduction show survivorship curves of the Deevey type III, where selective pressures are stronger in the seedling phase. Populations maintaining their numbers by vegetative propagation typically show an exponential rate of mortality (Deevey type II) (Sarukhán 1976). Thus, it is important that studies of survivorship examine genets, not merely ramets.

The fecundity of perennial plants also becomes difficult to measure unless it is simply the output per ramet. However, the genet may have a very large reproductive output, even though the number per ramet is few (Harper & White 1974). This is an obviously difficult problem to deal with in most populations.

The measurement of the relative energy investment in vegetative and seed reproduction is not as simple as it might at first appear. The vegetative structures that produce ramets are not specialized for only that purpose. A rhizome, for example, might function for absorption, anchorage, storage and propagation. A stolon may be partially supportive and photosynthetic while producing ramets. There is also considerable translocation of materials to and from structures involved in vegetative reproduction. Thus, biomass allocated to these structures is not strictly comparable to biomass allocated to sexual reproduction. In addition, calories and nutrients are allocated to structures of vegetative and sexual reproduction differently than biomass. In a study of five old-field goldenrods (*Solidago* spp.), it was found that nutrients (N, P, K, Mg, Ca) and calories were not allocated to rhizomes in the same proportions as to inflorescences (Abrahamson & Holler, unpublished data). It is difficult to equate vegetative and sexual reproduction on an energy cost basis. It follows that the summation of allocation to vegetative and seed reproduction is probably not an accurate measure of the cost of total reproduction.

As Sarukhán suggested, 'it becomes clear that an uncritical adoption of methods and concepts of animal demography to the study of plant populations is not wise'. Plant populations must be studied differently. The reader is referred to Sarukhán's studies of buttercups for an excellent approach (Sarukhán 1974; Sarukhán 1976; Sarukhán & Gadgil 1974; Sarukhán & Harper 1973).

Chapter 6
The Biology of Seeds in the Soil

ROBERT COOK

The focus of this chapter will be on the demography of seeds in the soil, that is, on changes in the numbers of seeds over time. The approach is in part demographic, being concerned with the effect of the number of seeds in the soil on the number of plants being recruited into populations; and in part evolutionary, asking how relative changes in the numbers of different kinds of seeds have in the past shaped the biology of species. I will present what the literature of seeds has to say on their demography in the form of several general statements, and I shall conclude that the most important demographic property of all seeds is the great variation in their capacity to remain dormant and viable in the soil. I will argue that this is a product of natural selection, i.e. it is adaptive, and forms a consistent part of the variation in the biologies of these species. This hypothesis will in turn suggest some avenues for further work to fill in very large gaps in our knowledge.

These gaps stem largely from the fact that much of the motivation for studying relevant parts of the biology of seeds has come from farmers whose interests have been primarily economical. First, farmers have been interested in storing a portion of their harvest for furture planting and have, therefore, wished to understand the nature of longevity in seeds. Secondly, farmers have always had to contend with the competitive effects of weeds among their crop plants and have been interested in where weeds come from and whether they might be destroyed while still seeds in the soil. Since the goals of farmers are primarily economical, such agriculturally oriented research introduces two possible biases into our review, namely that the species investigated have been either weeds or highly selected crop species, and the habitats in which they have been examined are usually the storage shelf or the garden. If one's interest is at all evolutionary, such biases must be considered in interpreting the characteristics of seeds.

DISPERSAL

There is wide variation between species in the capacity for seed dispersal, but most seeds enter the soil very near the parent plant.

The major emphasis in the classical study of seed dispersal and related aspects of seed and fruit biology has been concerned with questions of biogeography and the migration and assemblage of floras (Ridley 1930; van der Pijl 1969). Inferences concerning the nature and effectiveness of dispersal have been based on the study of fruit and seed morphology (Stebbins 1974) with the implicit asumption that the presence of obvious dispersal mechanisms is the product of continuing selection for the recruitment of offspring at some distance from the parent. This evolutionary and biogeographic orientation has therefore tended to emphasize the long-distance transport of propagules and the colonization of distant lands, although the actual recruitment of new genotypes is seldom observed. Yet recently there has been a shift to a more quantitative assessment of dispersal and recruitment growing out of an awareness that, in addition to the colonization of new habitats and the extension of the species' range, seed dispersal is most often involved in the replacement of the parent genotype upon its death. This transition can be seen in the questions raised concerning the effectiveness of achene-pappus dispersal in Composites (Carlquist 1967; Sheldon & Burrows 1973; Sheldon 1974). Carlquist argues that airborne dispersal has been of little significance in the colonization of the Hawaiian islands and, in addition, such colonizations are probably far more dependent upon the ecological conditions for establishment than upon the frequency of propagule arrivals. Sheldon concludes that much achene-pappus dispersal is very local, and ineffective for long distance dispersal, and that the pappus may have considerable selective importance in the germination and establishment of the seedling in relation to soil microsites.

There have been three additional steps in this transition from a species to a population orientation. Baker & Stebbins (1965) edited a set of essays from a symposium that brought together geneticists, evolutionists and population ecologists to consider precisely what was distinctive about the biology of colonizing species. It was quickly established that weeds are exceptionally good colonizers, and their capacity for dispersal is one important trait contributing to their success. Thus, a highly evolved capacity for dispersal became associated with a set of biological attributes or strategy, i.e. a large number of small seeds, rapid growth and development, broad habitat tolerances and developmental plasticity (Baker 1974).

Quantitative interest in dispersal also expressed itself in the work of Janzen (1970) who described a theoretical model to explain the diversity of tropical trees in relation to their species-specific seed predators. The dispersal of the seeds of any given adult will create a 'seed shadow' that indicates the proportion of the dispersal crop at any distance from the plant. In addition host-specific seed predators create a probability surface for survival of the seeds which increases at greater distances from the plant harbouring the predators. The product of these two functions is a population recruitment surface indicating the most likely distance between neighbours of the same species. There has

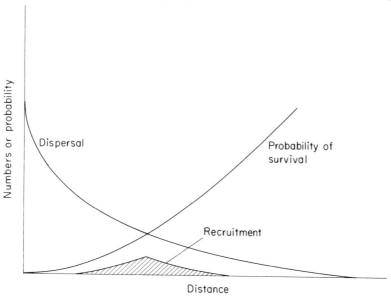

FIGURE 6.1. The recruitment of new genotypes as a function of the numbers of dispersed seeds and the probability of juvenile survival.

been some evidence to support these concepts (Janzen 1971a) but it has proven extremely difficult to get sufficient field data without complicating factors (Janzen 1975b). While the model only speaks to predatory sources of mortality of seeds and seedlings, the important demographic concept is that the probability of recruiting a new genotype into the population at any point in space is seen as a function of the number of seeds arriving and the probability of their surviving to maturity (Figure 6.1).

A final line of research that contributed to this shift began with the work of Salisbury (1942, 1974, 1975) and has continued in the research of Harper, Lovell & Moore (1970) and Baker (1972). In a survey of the seed size of many species of British plants in relation to habitat, Salisbury demonstrated that species whose seeds generally germinate in shaded conditions and species found in later phases of succession have larger seeds. This he related to the requirement of greater provisions for offspring growing at lower light levels and he concluded that 'the capacity to colonize in the face of competition appears to be associated with the amount of food reserve which the seed contains' (Salisbury 1942, p. 230). Harper *et al.* (1970) further suggested that all species not found in climax vegetation are continually doomed to local extinction and therefore must be able to colonize new terrain. Consequently selection will favour large-seed number, small-seed size and high dispersibility among such fugitive species, while favouring fewer, larger seeds (and, consequently, lower dispersibility) among species of shaded and climax vegetation.

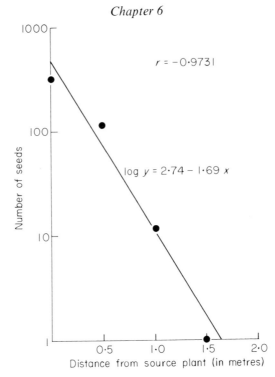

FIGURE 6.2. Pattern of seed deposition in teasel, *Dipsacus sylvestris*. The number of seeds deposited as a function of distance from the parent plant exhibits a negative exponential function (Werner 1975).

Thus, the small seeds of weeds and early successional herbs were seen to be the product of selection for colonizing ability out of the implicit assumption that the distance a seed disperses will be inversely related to its weight.

Direct evidence supporting any of these ideas is very meagre because of the difficulty in mapping the seed shadows of species. I know of only a handful of published dispersal curves in which the distribution of a cohort of seeds about a parent is portrayed (see Beattie & Lyons 1975; Howe & Primack 1975; Janzen *et al.* 1976; Rollins & Solbrig 1973; Salisbury 1942, 1961; Werner 1975b) in addition to those referred to in Levin & Kerster (1974). All of these indicate that most seeds do not disperse very far, and that past concepts of dispersal have tended to emphasize the long-distance tails of such distributions. Most distributions are quite skewed away from the parent plant and Werner (1975a) has suggested that a negative exponential function can provide a reasonably simple description of such distributions. Thus:

$$N_s = C \exp(-dX)$$

where N_s is the number of seeds, X is distance from the parent plant, C is a constant, and *d* is a measure of dispersibility under field conditions (Figure 6.2). That this measure probably differs even among closely related species is indicated by the work of Sheldon & Burrows (1973) in which the maximum dispersal distance for some selected composites in a 10 km non-turbulent wind was calculated to vary from 0·98 m for *Carlina vulgaris* to 7·57 for *Circium arvense;* however the bulk of the seed would fall close to the parent plant.

The demographic consequence of this distribution is that the effectiveness of dispersal depends upon the competitive influence of the maternal plant upon the recruitment of its offspring. If seeds germinate, grow and mature completely independent of their parent, then most new genotypes would be added to the population beside the maternal genotype where the probability recruitment surface is highest. This shifts the focus of interest in the study of dispersal to the mortality factors affecting the recruitment of offspring and their degree of association with the parent plant.

NUMBERS OF SEEDS

There are large numbers of seeds lying dormant in most soils, and these numbers depend upon the history of the surface vegetation and the age of the soil flora.

> I do not belive that botanists are aware how charged the mud of ponds is with seeds; I have tried several little experiments, but will here give only the most striking case: I took in February three tablespoonfuls of mud from three different points, beneath water, on the edge of a little pond; this mud when dried weighed only six and three-fourth ounces; I kept it covered up in my study for six months, pulling up and counting each plant as it grew; the plants were of many kinds, and were altogether 537 in number; and yet the viscid mud was all contained in a breakfast cup! (Darwin 1859).

Although Darwin may have been the first ecologist to sample the flora in soils, the bulk of information concerning the numbers of dormant seeds has come from the work of researchers interested in the control of weeds in cultivated soils (for references, see Jensen 1969; Kropac 1966; Roberts 1970). It was quickly discovered that the numbers of seeds can be very large indeed, completely dwarfing the number of plants at the surface. For instance Chancellor (1966) found that a dense stand of *Matricaria recutita* seedlings (280/m^2) was but 4 per cent of the number of viable seeds of that species in the soil. With many weed species easily capable of producing $2–5 \times 10^4$ seeds per plant (Salisbury 1942), numbers of seeds can be rapidly built up.

It was also noted that the number of viable seeds generally decreased with the length of time since the soil surface was last cropped, indicating a decay of seed numbers in habitats undergoing vegetational change. Early efforts to examine the seed flora of formerly cultivated soils (Chippindale & Milton 1934) yielded large numbers of weeds and other species not found as plants at

TABLE 6.1. The numbers of seeds in soils

Habitat types	Number of seeds/m^2	Source
Arable land	34–000–75 000	Brenchley & Warington 1930, 1933
Annual grassland	9 000–54 000	Major & Pyott 1966
Pasture	2 000–17 000	Champness & Morris 1948
Early successional fields	1 200–13 200	Oosting & Humphreys 1940; Livingstone & Allessio 1968
Tropical crop fields	7 600	Kellman 1974b
Tropical secondary forest (5 years old)	1 900–3 900	Kellman 1974b
Tropical rain forest	170–900	Kellman 1974b
Prairie	300–800	Lippert & Hopkins 1950
Forest Stands (80–200 + years old)	200–3 300	Oosting & Humphreys 1940; Livingstone & Allessio 1968; Kellman 1974a

the surface; samples of the seed flora beneath a diverse array of vegetations have also revealed high numbers of seeds with a negative correlation with age (Guevara & Gómes-Pompa 1972; Kellman 1974a, b; Livingstone & Allessio 1968; Major & Pyott 1966; van der Valk & Davis 1976). There is also an indication of a latitudinal decrease in the number of seeds in the soil (Johnson 1975), but this may just reflect changes in the composition of the flora with latitude rather than intrinsic properties of soils at lower latitudes.

An indication of the wide range of numbers of seeds in the soil can be obtained from Table 6.1. It should be noted that these numbers are very general since different samples show wide variation and accurate counting within a sample depends upon the method (germination or flotation separation) and the determination of viability. However the table does illustrate the important demographic generalization.

LONGEVITY

There exists great variation between species in the lifespans of their seeds, and some species have evolved the capacity to remain viable for centuries.

Interest in the viability and longevity of seeds has derived primarily from seedsmen and, more recently, plant breeders developing systems for the long-term storage of genotypes (Harrington 1972; Roberts 1972a). This has led to considerable theoretical explorations of the patterns of loss of viability under different environmental conditions and the ageing of seeds, as well as experimental approaches to the variability and control of viability. However, the primary focus has been loss of viability during storage on the shelf, and it will be important to distinguish this from the longevity of seeds while lying dormant in the soil (Villiers 1973), a condition far more relevant to the present essay.

The great variability between species in the longevity of their seeds has been revealed by two different approaches. The first involved the burial of seeds of a number of speceis in containers to be sampled at successive time intervals to detemine the identity of survivors. Thus, in 1879 Dr W. J. Beal began such a long-term experiment (Kivilaan & Bandurski 1973) by burying seeds of a number of common species in sand-containing, half-pint glass bottles, and another experiment was begun by Duvel in 1902 (Toole & Brown 1946). J. Lewis (1973) conducted a similar experiment and found that crop and grass species succumbed rapidly while legume and, particularly, weed and ruderal species remained viable for the entire period. This conclusion was supported by the periodic findings of the earlier experiments.

Secondly, seeds of species not present in the surface vegetation have been found in soil that has remained undisturbed for long periods of time. In one of the earliest efforts to quantify the number of viable seeds, Brenchley (1918) examined soils beneath pastures of known age since previous cropping and found that large numbers of weed seeds lay beneath pastures as old as 58 years. Other such studies (Champness & Morris 1948; Chippindale & Milton 1934; Hayashi & Numata 1971; Jalloq 1975; Lippert & Hopkins 1950; Livingstone & Allessio 1968; Major & Pyott 1966) have all found seeds, particularly of ruderal and weed species, beneath vegetation of great age, such as mature forests and pastures.

The greatest longevities have been reported by Ødum (1965, 1974) from archeological diggings and the ruderal soils beneath human habitations of known age. Many of the species found were annuals and short-lived perennials, and agricultural weeds predominated. For instance, *Chenopodium album* and *Spergula arvensis* were discovered in freshly cut soil known to be over 1 700 years of age. Ødum (1974) constructs a list of 27 species whose seeds continue to accumulate in soil because of their general abundance or effective wind dispersal, and a second list of 73 species whose presence is due mainly to the longevity of their seeds; 27 species have minimum ages over 100 years. Perhaps the greatest longevity record is that of *Lupinus arcticus* whose frozen but viable seeds were reported to be over 10 000 years old (Porsild, Harington & Mulligan 1967), although this finding could not be duplicated (Kjoller & Ødum 1971) and was called into question, along with all such findings, by Godwin (1968) because the purported age was based only on association with archeological material rather than direct dating. Despite this it seems clear that some species such as *Chenopodium album* have evolved a tremendous capacity to remain viable while buried. This prompts one to ask how much of the life of such weedy 'annuals' is normally spent lying dormant in the soil habitat. Table 6.2 gives an indication of some of the reported longevities of seeds that have been buried in the soil.

Chapter 6

There is a great wealth of information on the longevity of seeds kept in storage (Harrington 1972); and there is again wide variation between species, with most aquatic species, tree species that form nuts, some tropical species (generally economic) and crop species having short-lived seeds, and many short-lived weedy and ruderal species having long-lived seeds. In most cases the storage life was considerably shorter than the soil life, but it is very unclear what the relation is between the environment ideal for storage conditions and the environment of the soil.

A number of experiments have shown that the storage environment, particularly temperature and moisture, does greatly affect seed viability, and Roberts (1972a) was able to express much of the data in an equation:

$$\log \bar{p} = K - C_1 m - C_2 t$$

where $\bar{p}$ is the mean viability period, t is the temperature in degrees centigrade, m is the moisture as a percentage and K and C are constants. Clearly low temperatures and moistures favour viability, and Roberts also points out that decreasing the oxygen pressure will also increase longevity. A second important fact about the loss of viability in storage is that the distribution of deaths in time is described by a normal distribution such that seed survival curves under constant conditions are negative cumulative normal distributions (Figure 6.3). Roberts (1972b) discusses such curves in terms of a model involving a critical number of cells that must all suffer randomly distributed death before the seed becomes inviable. While this may prove important to

TABLE 6.2. Seed longevities in soil and decay rates (g) of populations

Species	Longevity (years)	Decay rate (g)
Chenopodium album	±1700	0·105
Thlapsi arvense	30	0·122
Polygonum aviculare	±400	0·156
Viola arvense	±400	0·161
Fumaria officinalis	±600	0·195
Eurhorbia helioscopia	68	0·206
Poa annua	±68	0·237
Capsella bursa-pastoris	35	0·244
Stellaria media	±600	0·252
Papaver rhoes	26	0·260
Vicia hirsuta	25	0·305
Medicago lupulina	26	0·340
Senecio vulgaris	±58	0·340
Spergula arvensis	±1700	0·340
Ranunculus bulbosus	±51	—
R. repens	±600	—

Longevities from Harrington (1972) and decay rates from Roberts & Feast (1972).

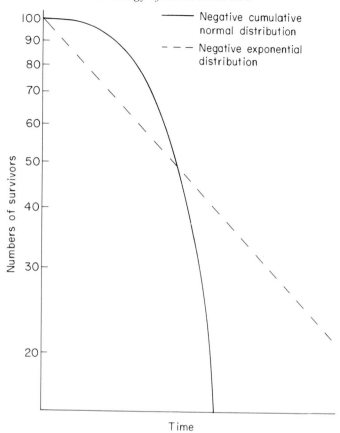

FIGURE 6.3. The negative cumulative normal distribution of seed viability during shelf storage and the negative exponential distribution of survival of seed in the soil.

problems of seed storage, its relevance to seeds in the soil is unclear. It focuses attention on the sources of mortality of seeds in the soil that have presumably been the selective forces adapting some species for great longevity.

It has been pointed out (Villiers 1973; Wareing 1966) that seeds in the soil display a much slower loss of viability than air-dry stored seeds, and that one significant difference between the two conditions is that many seeds in natural habitats will be dormant while fully imbibed with water. Villiers (1972, 1974) has noted that an important source of mortality in stored seeds is the accumulation of damage to cytoplasmic organelle membranes, as well as deleterious effects to the nuclear chromosomes. He further argues that seeds which are fully imbibed but prevented from growth by dormancy mechansims are quite capable of metabolic activity (food interconversions, membrane synthesis, organelle production) and are able to repair membrane and nuclear DNA damage as it occurs. Thus, viability in the soil is greatly extended by the normal functioning of enzyme repair systems that cannot operate in seeds whose moisture content has been reduced to a level that prevents germination during storage. Thus the longevity of seeds in the soil depends critically upon the dormancy mechanisms that prevent fully imbibed seeds from germinating, and it might be suspected that the pattern of mortality of seeds in the soil will be related to the breakdown of these dormancy mechanisms rather than the normally distributed loss of viability displayed by seeds in storage.

MORTALITY

Seeds living in the soil suffer a constant probability of death which differs between species.

The evidence for this rather sweeping and demographically important conclusion is, at best, limited. It derives from the seed burial and sampling experiments of H. A. Roberts examining the decay rates of populations of weed seeds living amidst different degrees of soil disturbance (Roberts 1970; Roberts & Feast 1972, 1973). In one series of experiments the investigators buried samples of twenty weed species in soil-containing cylinders and, over the course of 6 years, they sampled the cylinders to determine the percentage viable and noted the germination of any seedlings. In another experiment, garden plots were sampled over 5 years to determine the numbers of weed seeds in the soil; no new seeds were allowed to enter the soil. It was found that the numbers of seeds decreased exponentially such that the percentage loss each year was constant, with the rate of loss considerably higher in soils that were disturbed by frequent mixing. This implies that the probability of loss is constant, and the changes in numbers may be described by a negative exponential equation (Roberts 1972c):

$$N = N_0 \exp(-gt)$$

where N is the number of survivors at time t out of an initial population of N_0 and g is a constant for any set of conditions expressing the decay rate of the population. In addition the value of this decay rate appears to be quite different for different species (Table 6.2). For instance *Chenopodium album*, a species noted for its ability to persist in the soil, had a decay rate of 0·105 such that 53 per cent of the population of seeds survived for 6 years in undisturbed soil. Only 24 per cent ($g = 0·237$) of the population of *Poa annua* survived 6 years. In Figure 6.3 a hypothetical decay curve is plotted on a semi-log plot such that it becomes a straight line with a slope equal to $-g$. For the purposes of comparison I have also drawn a negative cumulative normal distribution, representing the loss of viability of seeds in storage, with a mean viability period equal to the half-life of the population of seeds in the soil. Clearly the juxtaposition of the curves depends upon the values of the constants chosen, but the differences in shape suggest that many seeds in the soil suffer mortality before the loss of viability while some have viability considerably enhanced.

In the experiments of Roberts & Feast (1972) less than 10 per cent of the loss of seeds from the population was due to germination and emergence of seedlings at the surface. Based on what little evidence there is (Taylorson 1970; Schafer & Chilcote 1970) it appears that most mortality is due to the breakdown of dormancy mechanisms and subsequent germination while buried in the soil (Roberts 1972c), although it should be pointed out that loss due to pathogens is very difficult to rule out. This implies that much of the variation in the decay rate of populations of different species (i.e. the value of g) is due to differences in the mechanisms of dormancy that have been the product of selection for the ability to persist in the soil.

DORMANCY

Small seeds survive for longer periods in the soil and their dormancy is frequently associated with a requirement for light.

Both the physiological and ecological literature on dormancy in seeds is vast (Amen 1974; Roberts 1972c; Taylorson & Hendricks 1977; Villiers 1972) and I shall try to extract that selected portion of it that seems most relevant to this essay. Because much of the research on the great variety of dormancy mechanisms has derived from economic and physiological interests, the classification of types of dormancy has concentrated on morphological and physiological properties of the seeds (Villiers 1972). Harper (1959) has suggested that 'some seeds are born dormant, some achieve dormancy and some have dormancy thrust upon them', and thus he has established three categories of dormancy—innate, induced and enforced—which were somewhat more ecologically oriented than previous classifications. Innate dormancy describes the condition of the seed while on the mother plant and during dispersal that prevents viviparous germination of the embryo. Nearly all plants display some form of innate dormancy and there is usually considerable variation in the duration of this dormancy on a given plant. Sometime after dispersal a seed may have its innate dormancy replaced by induced dormancy wherein the seed may be imbibed but germination is prevented by some inhibitory factor of the environment. This type of dormancy is capable of persisting even after the removal of the external conditions because an internal dormancy has been induced. The failure of seeds to germinate during enforced dormancy is due to some direct limitation of the environment which, upon removal, leads to germination. It is this latter dormancy which describes the condition of most seeds buried in the soil (Roberts 1972c).

Concern with the relation between dormancy and the fate of seeds in the soil at any point in time has led to efforts to partition seeds among a set of mutually exclusive categories. Roberts (1972c), after Schafer & Chilcote (1969), has suggested that:

$$S = P_{inn} + P_{ind} + P_{enf} + D_{gd} + D_{ge} + D_{ni} + D_{na} + D_{np}$$

where S = the total seed population, P_{inn} is the fraction of seeds in innate dormancy, P_{ind} = induced dormancy, P_{enf} = enforced dormancy, D_{ge} is the loss due to germination and emergence as a plant, D_{gd} = germination and death in the soil, D_{ni} = initially inviable, D_{na} = loss of viable due to physiological ageing and D_{np} = loss of viability due to predators or pathogens. Because of the difficulty of recovering seeds in the soil, few attempts have been made to partition an entire population of seeds into all of these classes. Schafer & Chilcote (1970) found that most of the loss (85 per cent) from a population of *Lolium perenne* was due to deep germination resulting in death (D_{gd}). In the same study they found that a greater proportion of the seeds of *L. multiflorum* was in induced (7 per cent) or enforced (30 per cent) dormancy while only 49 per cent germinated deep, a result compatible with the very weedy nature of the latter species. However, it should be noted that seeds in these experiments were buried in Saran mesh bags which may well have influenced their behaviour.

Of considerably more demographic relevance is the study by Sarukhán (1974) of the dynamics of sown seed of three species of *Ranunculus;* this was part of a larger comparative investigation of the population biology of *R. bulbosa, R. acris* and *R. repens* in grazed pastures over the course of 2 years. In April seeds of the three species were sown in circumscribed replicate plots and sampled six times over the course of 14 months; estimates of buried seed were made and no new production of seed allowed. The survivorship curves for the seed of the three species are shown in Figure 6.4. Although a great deal of sown seed was lost to vertebrate predators, it can be seen that *R. repens* has a much lower rate of decay than the other two species even though a higher proportion of its seed was consumed by birds and rodents. Most of the

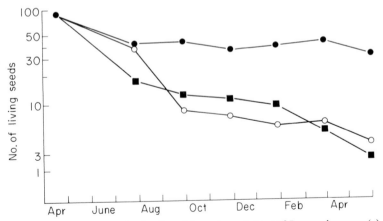

FIGURE 6.4. The decline of the living fraction of the seed populations of *Ranunculus repens* (●), *R. bulbosa* (○), and *R. acris* (■) in the field (Sarukhán 1974).

non-predatory losses from the *R. acris* and *R. bulbosa* populations were due to germination in April and September respectively, with little seed remaining dormant in the soil. This difference is quite compatible with the fact that most recruitment of new individuals of *R. acris* and *R. bulbosa* occurs from seed while *R. repens* depends upon vigorous asexual reproduction via stolons (Sarukhán & Harper 1973).

The role of seed size in the observed differences in the degree of dormancy is unclear. It is true that *R. bulbosa* has seeds nearly twice the weight of *R. repens* (3·1 mg *v.* 1·8 mg) but *R. acris* is the smallest of all (1·4 mg) (Harper 1957). Seed biologists have long noted that species whose seeds display great longevity in the soil tend to have small seeds (Chippindale and Milton 1934; Crocker 1938; Koller 1972; J. Lewis 1973). However, species with large seed productions are more likely to turn up in samples of old soil, particularly if viability period is normally distributed. For energetic reasons, species with a great reproductive capacity will tend to have small seeds. What one would like to know is whether the small seeds in a given cohort produced by a female have a greater probability of surviving in the soil than the large seeds from the same cohort. Halloran & Collins (1974) did find an inverse relation between seed weight and hardseededness; but the only experimental evidence that at all resembles natural conditions is that of Bhat (1973). He buried seeds of *Indigofera glandulosa* that had been sorted into size classes in soil-filled polythene bottles for 8 months and tested them for permeability and germination percentage. He found that larger seeds broke dormancy much more readily. However the conditions of his experiment were hardly ideal for a true test of the hypothesis. One might suspect that the relation between seed size, dormancy and longevity will depend upon the mechanism of dormancy.

In this regard a very important set of experiments were those conducted by Wesson and Wareing (1969a, b). It has been known for some time that a requirement for light is intimately involved in the dormancy of many species (Mayer and Poljakoff-Mayber 1975; Smith 1975), and its ecological significance was fully appreciated by Sauer & Struik (1964) who sampled soil beneath forests *at night* and found that the emergence of seedlings (mostly of weedy species) was greatly promoted when the samples were churned in the light rather than darkness before germination trials. Wesson & Wareing (1969a) essentially repeated this experiment, taking cores of soil beneath a 6-year-old pasture at night, discarding the top 2 cm, sorting and dividing the sample and germinating in light or darkness. There was very little germination in the darkness sample, while the soil in the light produced an abundance of weed seedlings. In addition they dug square holes (again at night) to varying depths in this same pasture and covered the holes with either glass or asbestos while leaving others uncovered. The numbers of seedlings that appeared after 5 weeks are shown in Figure 6.5, and it can be seen that germination depended totally on light. The most remarkable observation was that many of the species that appeared *(Chenopodium rubrum, Plantago lanceolata, Polygonum persicaria, Spergula arvensis, Stelleria media, Veronica persica* and others) displayed no light requirement as fresh seed. Wesson & Wareing (1969b) demonstrated that burial actually induced a light requirement in seeds which maintained them in a condtion of enforced dormancy as long as they remained buried.

It has long been known that the light-stimulated germination of seedlings involves the phytochrome system (Smith 1975). This is probably adaptive because it not only allows the seed to determine the degree of burial, but also to determine whether it lay beneath a canopy of green leaves that selectively filters the red spectrum out of white light and consequently inhibits germination when conditions would be unfavourable for seedlings (Smith 1972). The precise mechanism of dormancy release in light requiring seeds is unknown (Duke, Egley & Reger 1977; Taylorson & Hendricks 1977) and it seems clear that there are complex interactions with temperature and hormones (Smith 1975). It is known, for instance, that gibberellic acids will overcome dormancy in many light-requiring seeds (Taylorson & Hendricks 1977) and it might be supposed that the light provides the stimulus for the synthesis or release of this hormone which in turn initiates the hydrolysis of stored materials and the expansion of the embryonic cells. However, light and applied gibberellins interact synergistically, indicating at least two pathways leading to germination. It seems reasonable to conclude that increased levels of gibberellins are required by the embryo to begin the growth processes that expand the cells of the radicle through the seed coat.

It would seem appropriate to seek a relationship between seed size, light and dormancy among species which selection has moulded for long life in the soil, and *Chenopodium album* would appear to be exceptional in this regard, with individuals easily persisting in the soil for 500 years or more (Ødum 1965) and requiring light for successful germination. Karssen (1970a, 1976a) has explored the relation between dormancy and light in this species and his findings are very suggestive. He divides the process of germination into three stages: stage 1 involves the splitting of the outer testa layer near the radicle; stage 2 has the radicle expanded but still enclosed within the inner testa layer and the endosperm layer; in stage 3 the radicle has protruded through the inner testa layers. The critical distinction is between stages 2 and 3. In the former the seeds are capable of being dried and retaining viability for more than a year while the latter represents irreversible germination; and Karssen (1976b) argues that the two stages represent different steps of hormonal action. It has been postulated that germination involves the interaction of at least three hormones, gibberellin, cytokinin and abscisic acid (Khan 1975). Karssen (1976b) has shown that light and gibberellin are effective during stage 1, the splitting of the outer testa, while cytokinins and abscissic acid display antagonistic effects during stage 3, the elongation of the radicle through the inner tests (Karssen 1976a, b) when light is no longer required (Karssen 1970b). Light, therefore, seems to be important in enabling the embryonic

In general, the ecologically most simple systems are those in which the biological composition of the system and the environmental conditions are most closely controlled (Figure 7.1). In intensive agricultural production (e.g. glasshouse production) there is close control of both the physical environment and the biological populations (the crop, pests and pathogens). Conversely, in the most extensive agriculture (e.g. rangelands), there is little control of either the environmental factors or the biological populations.

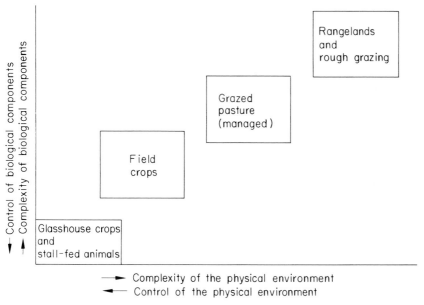

FIGURE 7.1. The relationship between control of the physical environment and control of the biological components of various agricultural systems (from Snaydon & Elston 1976).

DIVERSITY OF CROP SPECIES

Not only is there great diversity among agricultural ecosystems in the degree of control over environmental conditions and the biological populations, there is also great biological diversity among crop species and the pests and pathogens that invade them. This diversity has important implications in plant demographic studies. The diversity could be considered from an agricultural viewpoint, for example arable crops *v.* pastures, intensive *v.* extensive agriculture, but it is probably more relevant to consider it from a biological viewpoint, though the two classifications tend to coincide considerably.

Crops species vary in longevity from short-lived annuals, with life spans of as little as 10 weeks (e.g. barley), to long-lived perennials, with life spans of several centuries (e.g. some pasture grasses and tree crops). In some cases biennial species (e.g. sugar beet, cabbages) and perennial species (e.g. sugar cane, potatoes) are grown as annuals.

There are also large differences in the reproductive ability of agricultural species. On the one hand, many of the annual crops have been selected for greater seed production than their wild progenitors; on the other hand, many perennial crop species (e.g. some pasture grasses, sugar cane, bananas) are now unable to reproduce sexually and are reproduced only vegetatively. Crop species also vary in breeding system from apomixis (e.g. guayule) through inbreeding (e.g. wheat) to total outbreeding (e.g. maize). They also vary in genetic diversity from genetically uniform to highly heterogeneous.

This great diversity among crop species, together with the variation in the environmental conditions (from tropics to subarctic, from flooded rice paddy to semi-desert, etc.), and the variation in human control over both populations and environmental conditions, make it impossible to generalize about agricultural plant populations. For convenience, the diversity is subdivided here on the basis of lifespan of the species and the biological diversity within the stand, though great diversity still exists within each of these groupings.

UNIFORM ANNUAL CROPS

A large proportion of annual crops (e.g. wheat, barley) are genetically uniform, because they are inbreeding and have been intensively selected (Allard 1960). Other species, in spite of being naturally outbreeding, have been made uniform by intensive inbreeding and hybridization (e.g. hybrid maize). I include in this grouping biennial and perennial species which are used agriculturally as annuals. Some of these are reproduced clonally (e.g. potato, sugar cane) and are therefore genetically uniform, although naturally outbreeding.

As a result of these changes, most annual crops are genetically much more uniform than natural populations. In addition, the crops are usually more uniform phenotypically, as a result of management. For example, the seed or vegetative propagules are carefully graded and selected for uniformity, while the environment is modified to optimize conditions and reduce spatial variation. Perhaps the most important factor, however, is the synchronization of the life cycle of all plants by planting on a single day. These various factors together ensure that the crop stand is much more uniform than most natural stands, though some variation still exists.

A large proportion of the studies of crop populations have been concerned with density; this probably reflects the ease with which density can be manipulated and its importance in determining crop yield.

DENSITY AND THE INDIVIDUAL

Although the agriculturalist is usually more interested in the yield of the stand, rather than the performance of the individual plants, many studies of the effects of density have included studies of the effects on individual plants (e.g. Kira *et al.* 1953; Puckeridge & Donald 1967); these studies have been reviewed by Donald (1963) and Willey & Heath (1969).

Plant size and survival

Harper (1965) contrasted the effects of density on plant size (plastic response) and plant survival (mortality response). More recently the relationship between changing size and changing numbers has been investigated (White & Harper 1970; Harper 1977).

The relationship between density and plant size is similar for all crops and essentially the same for crops as for wild annual species already considered (Chapter 2). The many studies of the effects of density on plant size in crops have been reviewed by Donald (1963) and Willey & Heath (1969).

In agricultural practice, the density is rarely sufficient to cause appreciable mortality. This is partly because maximum crop yield is achieved at moderate densities (see below), but also because the uniformity of both the population and the environment (see above) ensure that few individuals are at a severe disadvantage. As a result, there have been few studies of the effect of density on plant mortality in crops. The studies that have been made (e.g. Donald 1963; Puckridge & Donald 1967; Yoda *et al.* 1963) show that mortality varies, depending on the species, the environment and the time after sowing. Mortality is greater, and occurs earlier, at higher densities (Figure 7.2b) but rarely

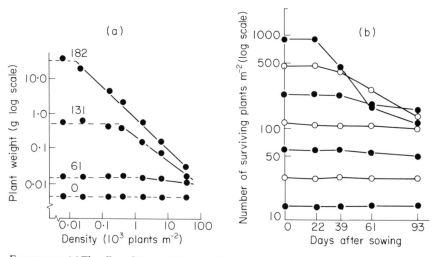

FIGURE 7.2. (a) The effect of density (plants m^{-2}) on log dry weight per plant (g) of *Trifolium subterraneum* at various times after sowing (data by Donald (1951), Figure by Willey & Heath (1969)). (b) The survival of *Glycine soja* plants at various times after sowing when sown at various densities. (data by Yoda *et al.* (1963), Figure by Harper (1977)).

occurs at densities less than 100 plants m^{-2}. At higher densities (10^4 plants m^{-2}) mortality does not occur in the first month but thereafter there is a steady exponential decay rate (Harper 1977; Yoda *et al.* 1963).

Interplant variation

Although annual crops are genetically and phenotypically uniform, compared with natural populations, phenotypic variation still occurs. Its effects have been studied. For example, Black (1958) investigated the effects of initial differences in seed size in a genetically uniform cultivar of *Trifolium subterraneum*. When large (10 mg) and small (4 mg) seed were sown separately, plant weight was similar after 90 days, when ceiling yield had been reached. Plant survival was initially greater from large seed but was slightly greater among plants derived from small seed. On the other hand, when the seed samples were mixed, plants derived from large seed were five times heavier than those derived from small seed after 110 days; all the plants derived from large seed survived but only 35 per cent of those from small seed survived. Unfortunately, Black did not study the effect on seed output, but Kaufman & McFadden (1960) did so with barley. They found that, when grown separately, plants derived from large seed (50 mg) produced 25 per cent more seeds than those derived from small seed (24 mg), though the seeds were slightly smaller. When the seeds were grown mixed, plants derived from large seed produced 70 per cent more seeds and the seeds were rather heavier than those produced by plants derived from small seed. The difference was mainly attributable to the number of fertile tiller per plant rather than the number of grains per tiller.

    Initial difference in chemical composition of seed, as well as seed size, may affect subsequent growth. For example, Bingham (1971) found that plants derived from seed of the same size, but containing 45 per cent more nitrogen, were 50 per cent larger after 23 days and 30 per cent larger after 60 days.

    Although the germination of most crop species is much more synchronous than that of natural populations, slight differences in time of germination still occur, primarily as a result of differences in depth of sowing. Black (1956) found that depth of sowing had little affect upon the initial seedling size and growth rate in a species *(T. subterraneum)* with epigeal germination. However, Arnott (1969) found very large effects of sowing depth on the emergence, initial seedling size and growth rate of a species *(Lolium perenne)* with hypogeal germination; the effect of planting depth was greatest for the smallest seeds. Differences in germination date are likely to be greatest in annual pasture species (e.g. *Trifolium subterraneum, Lolium rigidum, Stylosanthes* spp.) which retain dormancy mechanisms and regenerate naturally. Black & Wilkinson (1963) found that plants of *Trifolium subterraneum* that germinated 5 days later than surrounding plants were only 50 per cent as large as the others after 100 days, and those germinating 8 days later were only 25 per cent as large.

Several studies have been made of the frequency distribution of plant size in crop stands (Donald 1963). The initial seed size and early seedling weight is usually normally distributed (Koyama & Kira 1956; Naylor 1976; Obeid, Machin & Harper 1967) but, in most cases, the distribution soon becomes log-normal with large numbers of small plants and a few large plants (Koyama & Kira 1956; Obeid *et al.* 1967), though in some cases (e.g. Glenn & Daynard 1974), the distribution remains normal. In most cases the distribution becomes even more skewed than log-normal and, finally, a small proportion of plants contribute a disproportionately large part of the total stand yield and seed production, especially at high densities. Apparently the initially large individuals are more competitive and gain a disproportionately large share of the limiting resources at the expense of the smaller individuals. This is reflected in the fact that the yield of adjacent plants is negatively correlated (Donald 1963, Mead 1967). Little is known about the mortality of plants of different sizes or of the agricultural significance of variation in plant size, i.e. its effect on crop yield and quality.

Yield components

The way in which dry matter and chemical components are distributed within the plant is often of great agricultural importance; for example, the proportion of dry matter that occurs in the grain of cereals or the amount of sugar in sugar beet and sugar cane. Even in pastures, where the whole shoot is harvested, some components (e.g. leaf) have more dietary value than others. Increased density often affects the utilized component more severely than the total plant yield (Willey & Heath 1969). Bleasdale (1966) and Kira *et al.* (1953) found that the logarithmic relationship between plant weight and density (Fig. 7.2a) could be easily modified to describe the relationship for an individual plant component, since $W = K W_1 \alpha$ (where W = the weight per plant, W_1 = the weight of the plant component and K and α are constants). If $\alpha = 1$, then total plant weight and the component weight respond similarly to density. However, for most agricultural products (e.g. grain, edible root, sugar) $\alpha < 1$, so the component yield decreases more rapidly than total plant yield with increasing density.

Detailed studies have been made of the effects of density on the yield components of many crops (Puckridge & Donald 1967). At higher densities in cereal crops, the number of ears per plant, spikelets per ear and grains per spikelet are all reduced, though the weight of individual grains is usually less affected. There is considerable compensation among the various components. The number of tillers is especially variable, because of variation in both initiation (birth) and death. The numbers of spikelets per ear and grains per spikelet are also very variable and tend to compensate. Evans & Wardlaw

(1976) conclude that this compensation is one of the reasons for the success of cereals as crops.

The allocation of dry matter and chemical components is not only determined by density and other environmental factors, but is also determined genetically. Indeed, modification of this allocation appears to be much easier than modification of total herbage yield (Evans 1976) and has been a major factor in breeding for increased grain production. For example, the harvest index, i.e. the ratio of grain yield to total plant yield, has been progressively increased by breeders (Bingham 1969), though the total plant yield has slightly decreased. Similarly, the sugar content of beet has been greatly increased. However, compensation between various yield components (see above) usually prevents the increase of yield by selection for any single yield component.

The fact that selection for greater seed production has been so successful, while selection for total dry matter production or greater photosynthesis has been relatively unsuccessful (Evans 1976), seems to imply that there has been intense natural selection for total production but not seed production. It seems likely that, in the wild, a balance is struck between allocation of resources to vegetative production and seed production (Chapter 1). In crops, where there is essentially a single genotype stand and controlled density, more resources can safely be allocated to seed production. Indeed, there have been suggestions (Donald 1968) that the ratio of seed to total plant weight could be increased further, for example by the uniculum habit, and that breeders should select for low competitive ability. However, even in pure stands, seed production partly depends upon previous vegetative growth, since previous growth determines the number of fertile tillers (or branches) and the number of spikelets and florets (or flowers per axil). In mixed genotype stands (see below) vegetative growth is likely to be even more important, since it will partly determine competitive ability. Greater allocation of resources to seed usually leads to less allocation to vegetative growth; for example, in wheat the increased allocation of resources to the grain appears to be at the expense of root and tiller development (Evans 1976).

DENSITY AND CROP YIELD

Yield per unit area is of far more importance to the agriculturalist than yield per plant, so innumerable studies have been made of the effect of density on crop yield. The effects of genotype, management and environmental conditions on the relationship between density and crop yield have also been widely studied.

General relationships

Donald (1963), Holliday (1960) and Willey & Heath (1969) have reviewed the effects of density on crop yield. They concluded that total dry matter production usually increases asymptotically with increasing density (Figure 7.3), so does the yield of potato tubers and beet roots. On the other hand, the yield of grain in wheat, barley, maize, etc. decreases at high densities; this follows from the fact that grain yield per plant declines more rapidly than does total yield per plant (see above). As a result of these differences, the optimum density for grain or seed production per unit area usually occurs at a lower density and over a much narrower range of densities than that for total herbage yield.

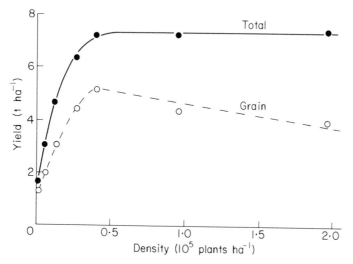

FIGURE 7.3. Typical effects of density on total stand yield and grain yield of maize (from Donald 1963).

Most crops differ from wild populations in that the plants are arranged in very regular patterns. Most crops are sown in rows, where the distance between individuals within the rows is much less than the distance between rows. In some cases the plants are sown in clumps or 'hills'. This regular distribution may effect the size of plants and the seed yield per unit area decreases as the rectangularity of the distribution increases, i.e. as the spacing within rows increases relative to the spacing between rows. Greater rectangularity also decreases the optimum density and makes the optimum sharper.

EFFECTS OF GENOTYPE AND ENVIRONMENT

The response to density and the optimum density, differs considerably between cultivars within many crop species (Figure 7.4a). Much of this variation has been the fortuitous outcome of breeding. On the one hand, selection at a particular density leads to a cultivar which performs best at that density

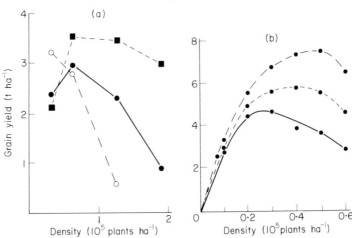

FIGURE 7.4. (a) The effect of density (plants ha^{-1}) on the grain yield (t ha^{-1}) of four contrasting cultivars of maize (Nelson & Ohlrogge 1957). (b) The effect of density (plants ha^{-1}) on the grain yield (t ha^{-1}) of maize grown with low (. . . .) medium (– – – –) and high (————) applications of nitrogen fertilizer (Lang *et al.* 1956).

(Russell 1969). On the other hand, selection for shorter stature or for upright leaves seems to have increased tolerance of high density in some cases (Anon. 1970) though not in others (Hicks & Stucker 1972). Cultivars of maize that are tolerant of high plant densities are more tolerant of shading (Stinson & Moss 1960) and seem to apportion a greater proportion of their resources to seed production (Moss & Stinson 1961), at least at high densities.

The pattern of response to density and the optimum density is also greatly affected by environmental conditions (Figure 7.4b). The optimum density and the maximum yield is usually increased when a limiting resource (e.g. N or water) is applied (Grimes & Musick 1960; Lang, Rendleton & Dungan 1956).

HETEROGENEOUS ANNUAL CROPS

Although most annual crops are genetically uniform, because of inbreeding and selection, some are genetically heterogeneous. A few annual crops are outbreeding (e.g. maize, rye), as are most biennials and perennials that are grown as annuals (e.g. sugar beet, carrots, potato); these crops are usually heterogeneous except if inbred and hybridized (e.g. maize) or intensively selected or vegetatively reproduced (e.g. potato). Even inbreeding crop species are usually heterogeneous if they have not been subjected to intense selection, so most 'land races' of inbred crops and semi-natural populations of inbred pasture species are heterogeneous. In addition to this essentially natural heterogeneity, artificial mixtures of cultivars have been used, both experimentally and in practice; reasons for this will be considered later.

The yield of the stand is most important from an agricultural viewpoint and the fate of the components is usually only of secondary interest. However, when viewed from a demographic viewpoint, the fate of the components is more important and so will be considered first.

GENETIC CHANGE IN HETEROGENEOUS POPULATIONS

Most studies of heterogeneous stands of crops have been based on experiments lasting only one season, i.e. one generation. It seems to be implicitly assumed by most agricultural scientists that the composition of the stand will be strictly controlled by sowing the true cultivar or mixture at the start of each season. This is a valid assumption for inbreeding crops in intensive agriculture but, in less intensive agriculture and in pastures, seed produced in one season normally forms the basis of the population in the next season. In this way cumulative changes in the genetic structure of populations can occur, often over many generations.

Single generations

Mixtures of genotypes or cultivars have been studied in many crop species, such as soybeans (Fehr 1973; Hanson & Hanson 1962), maize (Eberhart, Penny & Sprague 1964; Jurado-Tovar & Compton 1974), potatoes (Doney, Plaisted & Peterson 1966), wheat (Qualset 1968), flax (Khan, Putwain & Bradshaw 1975), groundnuts (Beg, Emery & Wynne 1975) and subterranean clover (Black 1963).

Many of the studies of interactions between genotypes and cultivars in mixed stands have been carried out to test the efficiency of selection methods in breeding programmes. Considerable genetic variation exists within families in the generations after crossing (F_2–F_6); selection in these generations must therefore be carried out in mixed stands; at first the variation is predominantly within rows and later between rows. Competition between genotypes and families might modify the performance of individual plants and make it difficult to select those plants that will ultimately yield most when grown in pure stands. There would be no problem if competitive ability was closely correlated with pure stand yield; indeed, competition might even exaggerate the differences and make high yielding genotypes easier to detect. However, most studies have shown that there is little correlation between grain yield in pure stands and mixed stands (Donald 1963). Some have argued (Allard 1961; Hamlin & Donald 1974) that there may even be a negative correlation, since the more competitive genotypes in mixed stands are vegetatively more vigorous and allocate less resources to seed production (see above).

Chapter 7

Sequential changes

The genetic changes in any one generation (or year) will accumulate if the change is consistent from generation to generation and if the seed of one generation is sown in the next. This rarely happens in intensive agriculture, but occurs in more primitive agriculture and has been studied experimentally in crop species (Harlan & Martini 1938; Suneson 1949) and in annual pasture species (Morley *et al.* 1962; Rossiter 1974). In most cases, cumulative changes lead to the dominance of one genotype (or cultivar), though the changes are not smooth, but fluctuate from year to year (Figure 7.5). Genotypes rarely disappear completely, even after 15–20 generations, but occur at low frequencies (< 5 per cent); probably due to frequency-dependent selection (see below).

The success of a genotype (or cultivar) in mixtures is determined first by environmental conditions. Harlan & Martini (1938) and Morley *et al.* (1962) found that different cultivars dominated mixtures at different sites (Figure 7.5b, c), while studies at a single site (Allard & Workman 1963; Rossiter 1974) showed that the relative success of genotypes varied from year to year (Figure 7.5).

The success of any genotype also depends in part upon the other genotypes in the mixture (Allard & Adams 1969). This success can usually be predicted, with reasonable accuracy, if the success of each genotype in mixtures with several others is known, i.e. competitive ability is general, rather than specific to some particular combination (Fehr 1973). However, success also depends upon the relative frequency of the genotypes. For example, Khan *et al.* (1975) found that cultivars of *Linum usitatissimum* were more competitive at low densities and *vice versa*. Such a relationship would lead to stable associations between cultivars (or genotypes). Evidence of frequency-dependent selection in crops, leading to stable associations of genotypes, has been obtained by Allard and Adams (1969) and Harding, Allard & Smeltzer (1966).

Few people have attempted to define which life cycle components determine success in mixtures. Morley *et al.* (1962) imply that flowering date, and hence seed production, in *Trifolium subterraneum* determined success. Jain & Marshall (1967) made a more intensive study in barley and also concluded that the early stages in development (germination and seedling establishment) were relatively unimportant, but that the number of seeds produced per plant was extremely important in determining success in mixtures. To some extent, of course, this is expected, since seed production is the integrated result of the various effects on the plant during its lifetime. Little is known about the attributes which affect competitive ability and ultimate seed production. Since success in mixtures is rarely correlated positively with seed yield in pure stands, and sometimes negatively correlated (see above), success seems to be determined more by vegetative attributes, such as height (Kern & Atkins 1970; Khalifa & Qualset 1975), leaf production or root production rather than by reproductive attributes *per se*. Differences in mortality seem to be less important than differences in plant size and seed production (Black 1961).

ever, be used with care because of the wide variation in the attributes and requirements of weed species and populations. Weed species vary from small herbs to shrubs, from ephemeral annuals to perennials, from prolific seeders to sterile plants, from parasites to autotrophs, and from apomicts to out-breeders; this variation is certainly as wide at that among the crops which they invade. We shall only consider annual species at this point.

DEMOGRAPHY OF WEED POPULATIONS

Surprisingly few studies have been made of the population dynamics of weed species, even though they provide an ideal opportunity to study relatively unrestricted population growth under relatively uniform and simple conditions. This again highlights the fact that agriculturalists are rarely interested in plant demography and population dynamics *per se,* though some of the data they collect may prove useful to the demographer and, conversely, some concepts in demography may perhaps prove useful in weed control.

A few studies have been made of the population dynamics of *Avena fatua* (Cussans 1976) and *Alopecurus myosuroides* (Naylor 1972). Selman (1970) found a more or less logarithmic increase in the density of *Avena fatua* growing in continuous barley. The annual rate of increase was approximately × 2·7 over a 6-year period. Wilson & Cussans (1975) found that the rate varied from × 1·6 to × 4·4, depending upon the type of cultivation and the treatment of straw before sowing. Herbicides must also have a large effect of the rate of increase, though no data seem to be published.

The most intensive experimental study of the population dynamics of a weed over a single season has been made by Harper & Gajic (1961) on *Agrostemma githago.* They studied the effects of weed density, and crop species and density, on weed establishment, mortality and seed production.

Although few comprehensive studies of population dynamics have been published, there are many published studies of some particular life-cycle components (Sagar & Mortimer 1976). The fate of seeds has been frequently and carefully studied (Brenchley & Warington 1930; Roberts 1968, 1970). The seed 'bank', in soils growing arable crops, ranges up to 10^5 m^{-2} (Roberts 1970). Most arable weed species have dormancy mechanism, so not all the seeds germinate in 1 year. If there is no further input of seed, the number of seeds declines logarithmically, due to death and germination (Roberts 1970). This reduction averages about 50 per cent per year, but depends upon the species, the depth of burial, the type of cultivation, the soil conditions (e.g. temperature and water content), and the amount of predation (Harper 1977; Roberts 1970; Sagar & Mortimer 1976). These various factors have not usually been controlled separately, so their relative effects are rarely known. Differences between species have been most commonly studied (Barton 1961), but then mostly under laboratory conditions, rather than in the field. Effects of temperature, depth and water content have also been studied in several weed species (Harper *et al.* 1970; Paterson, Boyd & Goodchild 1976; Thompson 1970). The differences between various species in response to these factors, and the agricultural significance of such differences, have rarely been studied.

The fate of weed seedlings has been less studied than the fate of seed. Sagar & Mortimer (1976) reviewed the available evidence and found that mortality ranged from 0 to 60 per cent. This mortality seems to be appreciably less than that of seedlings in nature and semi-natural communities; this presumably reflects the fact that weeds germinate almost synchronously with the crop and therefore grow in a fairly open habitat with less competition from established plants than exists in closed perennial stands. More weeds die later, when the stand closes; for example, Blackman & Templeman (1938) recorded mortalities of 76 and 86 per cent respectively for *Raphanus raphanistrum* and *Sinapis arvensis* after 9 weeks.

Seed output per plant varies considerably between annual weed species, though output also depends on environmental conditions. Sagar & Mortimer (1976) recorded differences ranging from 40 seeds per plant, for *Alopecurus myosuroides,* to 17 000 for *Papaver rhoeas.* Many of the seeds do not reach the soil seed bank. Many are carried off with the grain or straw, are destroyed by burning, eaten by animals or attacked by pathogens. For example Thurston (1964) estimated that 75 per cent of the seed of *Avena fatua* would be removed with the grain and 20 per cent with the straw, therefore if both grain and straw were removed, only 5 per cent of seed would be returned to the soil. However, if the crop was harvested late, 50 per cent of the seed might be shed before harvest (Wilson 1972).

Any weed seeds still present in the sown grain might form the foundation population of the weed in a previously weed-free area. Sagar & Mortimer (1976) calculated that only 1 per cent contamination of cereal seed by *A. fatua* would give 10 seeds m^{-2} and 1 per cent contamination by *Galium aperine* would give 50 seeds m^{-2}. This is easily sufficient to initiate a major infestation.

GENETIC STRUCTURE OF WEED POPULATIONS

Surprisingly few studies have been made of the genetic structure of populations of annual weeds. Excellent studies have been made within a few species, though many of those studies have been made in marginal agricultural areas, for example studies of *Avena spp.* by Allard and co-workers (Allard *et al.* 1968), studies of *Poa annua* (Ellis 1974; Law, Bradshaw & Putwain 1977) and studies of *Senecio vulgaris* (Abbot 1976).

In spite of the fact that most annual weeds are almost entirely inbreeding, considerable genetic variation has been found within most populations, though others were uniform (Jain & Rai 1974; Singh & Jain 1971). Variation within populations of inbreeders has been attributed to both genetic and ecological factors. On the one hand, even (< 5%) slight outbreeding (Allard *et al.* 1968) maintains variation, and there seems to be little relation between extent of inbreeding and amount of variation. On the other hand, environmental variation in space (Abbot 1976; Clegg & Allard 1972) and time, and the selective advantage of heterozygotes (Allard *et al.* 1968), can maintain variation within largely inbreeding populations. Long-term dormancy of seed may also maintain this variation.

It seems likely that the relative uniformity of the environment in agricultural crops (see above), might lead to greater genetic uniformity in weed populations. There is no real test of this, but circumstantial evidence (Jain & Rai 1974; Singh & Jain 1971) seems to indicate that variation is less under intensive agricultural management, but more information is needed. In addition, more studies are needed of genetic change in weed populations following changes in management (Hancock & Wilson 1976; Shontz & Shontz 1972) and the use of herbicides (Karim & Bradshaw 1968, Radosevich 1977). The few studies of the effects of herbicides on the genetic structure of plant populations contrast sharply with the abundant evidence of the effects of pesticides on insect populations (Georghiou 1972). In addition, further evidence is needed on the effects of pests and pathogens on the population dynamics and genetic structure of weed populations (Cussans 1974).

WEEDS AND CROP YIELD

Agriculturalists are less interested in the number of weeds *per se* than in the effect of those numbers on crop yield. Both surveys and experimental studies have been used to define relationships between weed density and crop yield.

Dew (1972) found negative correlations ($r = 0.67$ to $r = 0.99$) between crop yield and the square root of *Avena fatua* density. The size of the yield reduction depended on the crop. For example, 100 plants m^{-2} of *A. fatua* reduced the yield of barley by 22 per cent, of wheat by 33 per cent and of flax by 60 per cent. Similar relationships have been found for other weeds (Reeves 1976). Such general stochastic relationships may be useful for roughly estimating crop losses, but experimental studies show that the loss is also affected by crop density, fertilizer use and climatic conditions. Crop losses are usually less at higher crop densities (Felton 1976; Nieto & Staniforth 1961) since the crop suppresses the weeds. Nitrogen fertilizer also reduces crop losses (Nakoneshny & Friesen 1961; Nieto & Staniforth 1961; Scott & Wilcockson 1976) though this will only occur if N is the major limiting resource, and the result may sometimes be the reverse (Appleby, Olson & Colbert 1976; Scott & Wilcockson 1976). There have been surprisingly few studies to define the nature of competition between weeds and crops. Aspinall (1960) concluded that root competition was more important than shoot competition in mixtures of barley and *Polygonum lapathifolium*, but Schreiber (1967) found the reverse in mixtures of *Lotus corniculatus* and *Amaranthus retroflexus*.

Several studies have attempted to define 'critical periods' during the growth and development of the crop, when it is particularly susceptible to competition from weeds (Chancellor & Peters 1974; Glasgow, Dick & Hodgson 1976; Nieto, Brondo & Gonzales 1968; Roberts 1976; Scott & Wilcockson 1976). In some crops (e.g. onion, narcissus) there is a marked 'critical period'; early or late infestations have little effect on yield (Lawson & Wiseman 1978; Roberts 1976). In other crops (e.g. *Vicia faba*) there is no 'critical period' (Glasgow *et al.* 1976).

The effects of weeds on the yield components of the crop have been studied in several crop species. Weeds do not usually affect grain size in cereals and legumes (Felton 1976; Reeves 1976) or the number of grains per pod (Felton 1976) in legumes. Weeds do reduce the number of fertile tillers and the number of grains per ear in cereals (Reeves 1976) and the number of pods in legumes.

In summary, although annual weeds are ideal material for demographic studies and have provided useful items of information for the plant demographer, few studies of annual weeds have provided anything approaching a complete demographic census of population. In addition, apart from the comprehensive studies of *Avena spp.* by Allard and his associates, few studies have been made of the genetic demography of populations of annual weeds.

HETEROGENEOUS PERENNIAL CROPS

A wide diversity of perennial species are used in agriculture. Among herbaceous species, a few are used for fruit production (e.g. strawberry), and many are used in pastures. In addition, many perennial tree and shrub species are used in agriculture, these include species producing fruit (citrus, apples, coconut, grapes, *Ribes* spp, etc.), stimulants (tea, coffee, cocoa, grapes, etc.) and other miscellaneous products (rubber, oil palm, etc). Although most of these perennial crop species are outbreeding, not all are grown in genetically heterogeneous stands. For example, many of the tree and shrub species are reproduced clonally and therefore grown as genetically uniform stands. Some of the herbaceous perennials, for example strawberry, are also clonally reproduced. On the other hand, most pasture perennials are outbreeding and grown from seed; as a result stands are usually genetically heterogeneous. Some tropical pasture species are inbreeding, or even apomictic, but few have been so carefully selected that they are grown in genetically uniform stands.

Since most perennial tree and shrub crops are grown in genetically uniform stands, and are therefore more akin to homogeneous annual crops, this section will deal mainly with pasture species which are genetically heterogeneous.

The ecology, and hence the demography, of pasture plants differs from that of the annual crop species and weeds, and is more similar to that of wild species. In general, perennial pastures are ecologically more complex than annual crops. First, pastures are rarely grown in single species stands; Only a small proportion of pastures are sown with a single species, and even these soon revert to multispecies stands by the invasion of other species. As a result, only newly sown and experimental monospecific stands can be truly regarded as ecological populations. In all other cases, plants of any one species are more likely to interact ecologically with plants of other species, and must therefore be considered ecologically more in the context of the community than the population. Viewed from a genetic standpoint, the population may still be an entity, though it occurs within an ecological matrix of other species. Secondly, the physical environment of perennial pastures is less controlled and more variable, both in time and space, than that of annual crops. Thirdly, pastures are grazed by animals and this adds considerably to the ecological complexity. Perhaps the most important difference between annual crops and perennial pastures, from a demographic viewpoint, is the great importance of vegetative spread or reproduction in most pasture species, as opposed to seed production in most crop species. On the one hand, this leads to difficulties of defining a 'plant' for demographic purposes (Chapter 5); to some extent this has been overcome by using more easily identified plant units, such as tillers or stolons (Harper 1977; Jewiss 1972). On the other hand, the long-term survival of genotypes (or genets) will, under these conditions, usually depend more on vegetative reproduction than seed production.

In considering annual species, I considered crops and weeds separately. This is not possible with perennial pasture species, since there is a continuous spectrum from highly desirable species, which support the greatest animal output, to highly undesirable species, which are toxic and kill domestic animals. Between these two extremes there are many species which are often regarded as 'weeds', though many actually give as much animal production as the sown species. Other species may be as productive as sown species at a particular time of year, or in a particular environment, or may complement the sown species in the pasture, so that the mixture is more productive than the pure stand (Snaydon 1978b). In view of this, I shall consider all perennial pasture species together.

DENSITY AND STAND YIELD

Fewer studies have been made of the effects of plant density on perennial pasture yield than on crop yield. Density only affects pasture yields for a short time, because individual plants rapidly and stands reach the same ceiling yield within several months (Donald 1963; Kays & Harper 1974). Most of the studies that have been made (Kays & Harper 1974; Naylor 1976) have therefore been of the early stages of establishment. In contrast to this, however, pasture species are often planted and maintained at very low densities for assessment and selection in plant breeding programmes. Many studies have therefore been made of the performance of genotypes and cultivars under wide spacing (spaced plants) and close spacing (swards). Most have shown that there is only a slight correlation between yield in spaced plant conditions and in swards (Green & Eyles 1960; Knight 1960; Lazenby & Roger 1964); this result is similar to that for crops, where crop cultivars differ in response to density (Figure 7.4). However, others have found closer correlations, especially if the spacing is not too wide (England 1967; Lazenby & Roger 1964) or when the data are collected during the first 6 months after planting (Sedcole & Clements 1973).

DENSITY AND THE INDIVIDUAL PLANT

Increasing density reduces the size of individual plants, as in the case of annual crops (Figure 7.2a). The effect is due almost entirely to differences in tiller number rather than tiller size (Kay & Harper 1974; Naylor 1976). In both these studies, the number of tillers per unit area reached the same value, regardless of sowing density, within several months. The effects of density on plant (genet) mortality were similar to those for annual crops (Figure 7.2b); at the highest density (10^4 plants m^{-2}) plants began to die within 30 days and death progressed as an exponential decay rate thereafter; at the lowest density (320 plants m^{-2}) no appreciable mortality occurred for 120 days, but thereafter there was an exponential decay rate.

Longer-term studies of sown swards (Charles 1961; Langer *et al.* 1964) and of semi-natural pastures (Hawthorn & Cavers 1976; Hodgkinson 1976; Sarukhán & Harper 1973; Williams 1970) show that exponential decay rates continue over several years, though the initial rate may be greater than the subsequent rate. In addition, there are large seasonal fluctuations in the mortality rate and large differences due to grazing management and fertilizer applications. If exponential decay rates continued, with no replacement by seedlings, then the stand would soon be dominated by a few genotypes, presumably well adapted to the particular conditions. This appears to have happened in several long-lived perennial species, such as *Festuca rubra, F. ovina* and *Holcus mollis* (Harberd 1961, 1962, 1967) growing in semi-natural pasture. On the other hand, it should not happen in short-lived perennials, such as *Anthoxanthum odoratum* (Antonovics 1972) and *Ranuculus* spp. (Sarukhán & Harper 1973) where dead plants are replaced by young seedlings.

Individual plants or genets soon lose their individual identity in many pasture species. Plants spread vegetatively and intermix, while the new components (tillers or stolons) become separated from the 'parent' plant and become physiologically independent. Under these conditions it has become conventional, in agricultural studies, to measure and count the component units (e.g. tillers, stolons, leaves) rather than plants. Harper (1977) has also advocated the use of these components or units for demographic purposes (Chapter 3).

Several studies have been made of the demography and dynamics of grass tillers. Jewiss (1972) described several studies in which there were large seasonal variations in the origin and death of tillers. For example, most tillers of *Lolium perenne* and *Festuca pratensis* died in May, at the time of maximum dry matter production, and most new tillers appeared after this, in June and July (Langer *et al.* 1964). As a result of these seasonal patterns, the total tiller number was greatest in March–April, with a subsidiary peak in August–September. The number of tillers per plant increased, though somewhat erratically, throughout the 3 years of the study. The seasonal pattern of mortality of tillers in these studies was remarkably similar to that for whole plants in studies of several perennial pasture herbs (Cavers & Harper 1967; Sarukhán & Harper 1973). Apparently mortality is greatest at the time when growth is most rapid and therefore competition is more intense.

Apart from Naylor's (1976) study, there have been no studies of the frequency distribution of the size of individual plants, or any study of the size of plants that died. However, since the identity of plants of many pasture species is soon lost, such studies often only have meaning in the early stages of sward establishment. The frequency distribution of size in *tillers* also has little meaning since, at any one time, the tillers present differ in age from 1 day to 1 year (Jewiss 1972). The life history of these tillers has been followed (Jewiss 1972). Those originating in summer tend to have a high initial rate of mortality, but a low mortality rate over winter. Those originating in winter and spring have a low initial mortality but a high mortality in late spring; these differences seem to be due to seasonal conditions (see above), rather than the age or size of tillers.

SEED PRODUCTION

The factors affecting seed production of the more important temperate, agricultural grass species have been closely studied. For example Langer (1957) and Lambert (1964) studied the effects of cutting, density and nitrogenous fertilizer on the seed yield of *phleum pratense*. Progressively later cutting, after early April, decreased seed yield, nitrogenous fertilizer increased yield. The effect of density on seed yield is similar to that for annual species (Figure 7.3). Ryle (1964) has made a more detailed study of the seed production of individual tillers, especially in relation to date of origin and nitrogen supply. There was an almost linear relation between tiller age and seed production; tillers originating in October produced almost twice as much seed as those originating in the following March.

These detailed studies have been carried out on young, single-species stands under experimental conditions. Much less is known about seed production of grass species in old mixed-species stands under normal agricultural conditions. There is some evidence that the genotypes that survive in old pastures, probably as old plants, may produce little seed (Hayward & Breese 1966), but again little information has been obtained under field conditions.

THE FATE OF SEED AND SEEDLINGS

The seed content of soil beneath pastures has been less carefully studied than that of cultivated soils (see above). Studies, such as those of Chippindale & Milton (1934) and Milton (1936), have shown that up to 40 000 seeds m^{-2} can occur in the soil (Sagar & Mortimer 1976); about half of these occur in the top 3 cm since the soil is undisturbed (Chapter 6).

Most studies of the germination of seed, and the factors that control it, have been carried out either under controlled conditions (Harper *et al.* 1970; Heydecker 1973) or in newly sown pastures (McKell 1972). However, a series of recent studies have followed the fate of seed and seedlings of several herb species in agricultural grassland (Cavers & Harper 1967; Hawthorn & Cavers 1976; Putwain & Harper 1970; Sarukhán 1974; Sarukhán & Harper 1973; Thomas & Dale 1975). These studies showed that few seeds, whether naturally dispersed or sown, germinated and established. The actual percentage success depends heavily upon the life span of plants and their ability to spread vegetatively (Sarukhán 1974).

POPULATION DYNAMICS

There seem to have been no complete studies of the overall population dynamics of grass species in agricultural pastures, probably because of the extreme difficulty of identifying individual plants and identifying seedlings. However, the detailed studies of grassland herbs by Hawthorn & Cavers (1976), Sarukhán (1974), Sarukhán & Harper (1973) and Thomas & Dale (1975) have greatly increased our understanding of the population dynamics of non-grass species. These studies have already been considered in previous chapters and also by Harper (1977).

GENETIC CHANGE

So far we have largely considered the population of perennial plant species as genetically uniform, but this is a great oversimplification since most are outbreeding and contain considerable genetic variation.

Change with time

Since there is considerable genetic variation, and since the production of seed, the germination and establishment of seedlings and the growth of adult plants are all highly dependent upon genotype, the genetic structure of populations is very liable to change. Genetic change occurs within agricultural pasture populations both during establishment and after (Snaydon 1978a). It also occurs during seed multiplication, i.e. after release by the breeder but before use by the farmer (Snaydon 1978a).

Genetic changes occur during seed multiplication because of inherent

Chapter 8
Demographic Problems in Tropical Systems

JOSÉ SARUKHÁN*

It has been variously stated that plant demography, as a *corpus* of knowledge, is a newly born branch of plant ecology. If so, tropical plant demography (and I would also add tropical plant population ecology) is still very much in gestation. The strong emphasis on descriptive approaches such as morphological and phytosociological studies that characterize plant ecology in general is still more pronounced, at least relative to the amount of work done, in the so-called 'tropical' plant systems.

The task of collating and interpreting relatively large bodies of information about ecological phenomena to find general patterns becomes considerably hindered when one deals with studies on different topics of population ecology of tropical plants. One reason is that such information is remarkably scanty, particularly if one wants to use the information for its possible demographic relevance, even at a general level. Another reason is that a considerable amount of potentially useful observations with demographic applications, available in the tropical literature, is the result of forestry-management oriented studies, which often either fail to have ecologic insight, or are methodologically unrigorous, being therefore, at best, of very limited use. However, certain aspects of the life cycle of a plant (such as the seed stage, for example) have received considerably more attention than others and general patterns of behaviour and adaptive responses of plants have already been proposed, as will be discussed later.

In this chapter, I make reference to those topics of population ecology of tropical plants that are centrally relevant to population attributes and that can be interpreted from a demographic point of view. This will, in consequence, leave out large areas of ecological information such as morphological and physiological responses of plants to physical environmental factors, studies of energy flow in communities, much of what is known of floral biology *per se*, etc. I will also constrain this paper to the intertropical area which is clearly frostfree with 1 m or more of rain per year, whether of even or seasonal

* Parts of this paper were read by D. Piñero, who made valuable comments. Research reported in this paper has been aided financially by the Programa Nacional Indicativo de Ecologia Tropical, CONACYT. Mexico.

distribution. This leaves out a number of habitats which occur in the 'tropics' such as many conifer and broadleaved forests where a certain amount of demographic information is available.

In this chapter, I review the state of knowledge of plant demography and related population biology in the tropics, pinpointing some of the problems I consider to be important.

POPULATION STRUCTURE AND DYNAMICS: NUMERICAL PROPERTIES

POPULATION FLUX MODELS AND LIFE TABLES

All studies in the tropics on some aspect of plant demography deal with arboreal species. A few have lead to the elaboration of different flux models which, to a greater or lesser degree, cover the life cycle of the plants. There is a wide variation in the nature and detail of the information gathered. However, the time span of the studies is generally short. Most of them cover periods of between 1 and 3 years, occasionally more for certain restricted aspects of population ecology. This strongly limits comparison and discussion of the data for synthesizing purposes. In dealing with certain sets of data, one is often tempted to manipulate these in order to obtain more comparable groups of results, often imposing on such 'reanalyses' seriously crippling assumptions. As far as possible, I have resisted such temptation.

Perhaps, the first published formal approach to a demographic study of tropical plants is that of Fournier & Salas (1967), with a single cohort of seedlings of *Dipterodendron costaricense* germinated under a parent tree; observations lasted for a single year for which a life table was prepared. As the information pertains to seedling survivorship, it will be discussed later in the paper.

Bannister's (1970) study of *Euterpe globosa* (= *Prestoa montana*), expanded later into a more formal demographic analisis by Van Valen (1975), first showed that palms can be extremely suitable subjects for demographic studies. *Euterpe globosa* is a very common plant in the rain forest growing on slopes of mountainous islands in the West Indies. It is a canopy species up to 20 m in height. In the Luquillo Mountains in Puerto Rico, site of Bannister's study, it ranks second in density, basal area dominance and volume after *Dacryodes excelsa* (Briscoe & Wadsworth 1970; Smith 1970).

In the survivorship curve published by Bannister (1970) (Figure 8.1,) *Euterpe globosa* appears to be in a state of equilibrium. She suggests that palms face a very critical point at a height of some 6 m (*c.* 50 years of age), very probably due to the difficulty of maintaining a positive metabolic balance at this point. Palms which survive beyond this point increase dramatically their survivorship and may start reproducing. Detailed aspects of reproductive

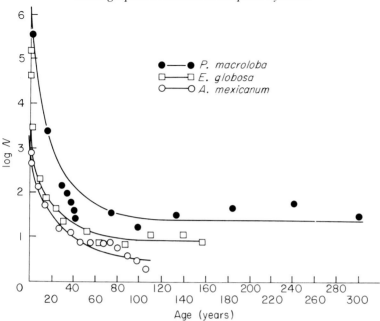

FIGURE 8.1 Survivorship curves for whole populations of *Astrocaryum mexicanum* (Sarukhán 1977) (o——o), *Pentaclethra macroloba* (calculated on data of Hartshorn (1972)) (●——●) and *Euterpe globosa* (adapted from Van Valen (1975)) (□——□).

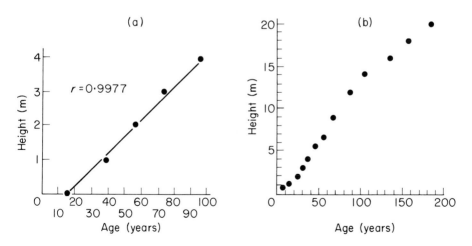

FIGURE 8.2. Size–age correlations for two species of tropical palms, (a) *Astrocaryum mexicanum* and (b) *Euterpe globosa* (data from Van Valen (1975)).

schedules were not taken into consideration in the study, as was the case for some aspects of rates of leaf reproduction and leaf-scar distance in plants of different ages, both necessary to estimate age of the individual palms (Figure 8.2b).

Hartshorn's (1972, 1975) studies on *Pentaclethra macroloba,* a dominat canopy species and *Stryphnodendron excelsum,* a contrasting, less abundant species in the tropical rain forest at Finca La Selva in Costa Rica, constitutes a first effort into analysing demographically tropical arboreal species and applying projection matrix models to the information on survival and growth of individual plants. The study lasted 12 months and Hartshorn's results show that population numbers and size distributions (since ages could not be detected) were essentially stable. Simulated increases in mortality in the model (i.e. increased seed predation) showed no drastic effects on the population size of *Pentaclethra macroloba.* This species differs from many other tree species typical of the dominant layer, in the remarkable abundance of individuals between the young seedlings and the emergent tree stages. As a general rule, all emergent tree species in species-rich rain forests notoriously lack recruitment between a very abundant and continuously renewing seedling population and the huge emergent individuals. *P. macroloba*'s situation fits in what Whitmore (1974, 1975), describes for South-east Asian forests as 'shade-bearing' species, whose seedlings and sapling can persist for long periods growing at very low rates. Size distribution of these species have a characteristic 'reverse-J' shape.

Pentaclethra shows the 'normal' or 'reversed-J' distributions of size classes (in this case a combination of heights for plants up to 3 m, and d.b.h. for taller trees) of what the foresters consider a 'stable or self-regenerating' stand. Applying the growth rates found by Hartshorn for the different categories of trees, I estimated the time a tree spends in each category, and hence individual age, assuming all trees in a category grow at the same rate. When plotting survivorship against age, a reasonable image emerges (Figure 8.1) which corroborates the suggestion that populations of *P. macroloba* possess a fair degree of stability and follow the survivorship curve generally found for sexually reproducing plant species, whether tropical or temperate, herbaceous or arboreal.

Of the few attempts at demographic studies in the tropics, one study has dealt with a species of extratropical origin (phytogeographically speaking) occurring naturally in areas of Central America. This is Hutchinson's (1976) study on *Pinus caribaea,* at the moment the most important timber tree being planted in the lowland tropics. An attempt was made at modelling population and individual growth of the species, and some interesting results on seed population dynamics were obtained. However, the extrapolation of growth rates and density interactions from temperate zone pine species to *P. caribaea* in this study strongly limits the discussion of the general demographic model obtained.

A series of papers have appeared (Vandermeer, 1977a,b,c; Vandermeer,

Stout & Miller 1974) on population ecology and demography of *Welfia georgii*, a common subcanopy palm of the lowland rain forests of eastern Costa Rica, although the final demographic analysis as such is still in preparation. These relate to growth rates of the palm, density-dependent survivorship of seedlings, seed dispersion and predation, and will be dealt with in the respective sections later on.

Sarukhán (1977) published a long-term comparative demographic study of both temperate and tropical aboreal species (*Pinus hartwegii, Cordia elaeagnoides, Nectandra ambigens* and *Astrocaryum mexicanum*) which belong to strongly contrasting environments (timberline single-species conifer forest, tropical evergreen and tropical deciduous forests). Within these species, the most complete set of demographic data has been obtained for the palm, *A. mexicanum*, typical of the lower strata of the tropical evergreen forests of southern Veracruz, Mexico, and often the most abundant arboreal species in the forest; it is this palm which imprints the characteristic physiognomy to the undergrowth since its maximum height is c. 6 m. This species has proven to be, as is the case with many other palms, an excellent subject for detailed demographic studies. An ample description of its biology, morphological characteristics, phenology, population structure, etc., has been published elsewhere (Piñero, Sarukhán & González 1977).

The particular growth habit of this species allowed, with the knowledge of leaf-production rates at different ages, to establish approximate individual ages, which in turn were found to be correlated fairly closely to the size (height) of the palms (Figure 8.2a). This offered, as in the case of *Euterpe globosa* the possibility of developing formal life table analysis.

TABLE 8.1. Survivorship probabilities for individuals of *Astrocaryum mexicanum* from seedlings to the oldest mature stages recorded in the population. Data are averages of six 600 m² permanent sites (Sarukhán 1977)

Stage	Age (yrs)	p to survive to the next stage
seedling	8	0·37
juvenile	15	0·32
immature	27	0·81
	39	0·64
mature	47·5	1·00
	56·0	1·00
	64·5	1·00
	73·0	0·75
	81·5	0·66
	90·0	0·75
	98·5	0·66
	107·0	—

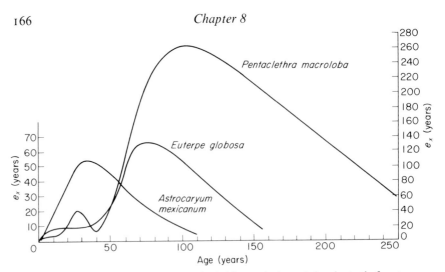

FIGURE 8.3. General trends of life expectancies (e_x) for tropical trees belonging to the forest undergrowth (*Astrocaryum mexicanum*) (calculated with data from Sarukhán (1977)) and to the emergent strata of the forest (*Euterpe globosa* (Van Valen 1975) and *Pentaclethra macroloba* (data calculated from Hartshorn (1972)).

Based on a vertical life table, a preliminary population flux model was prepared (Sarukhán, 1977) for six permanent sites, representing different densities of the palm, including information about numbers at each age (or size) class, the probabilities of survival to each subsequent category, reproductive contributions of each category, etc. Table 8.1 shows survival probabilities from seedlings to the oldest categories. Clearly, the strongest reduction of numbers of plants occur at the passage from seedlings to juveniles and from these to the young immature (non-reproducing) palms. After this stage, chances of a plant remaining alive are rather similar until the oldest recorded sizes or ages in the populations are attained.

The former is more clearly seen in Figure 8.1 which represents the survivorship curve of *Astrocaryum mexicanum* individuals which is essentially identical to that of *Euterpe globosa* and *Pentaclethra macroloba*. The drop in survivorship at the late stages both in *Astrocaryum mexicanum* and *Euterpe globosa* could be due to the effect of the small numbers of plants observed at these ages.

LIFE EXPECTANCIES

An interesting analysis of data from life tables which use age as a pivotal axis is the comparison of the life expectancies (e_x) of species with different life habits. Reanalysing data of Van Valen (1975) for *E. globosa* and comparing it with data for *Astrocaryum mexicanum* from Sarukhán (1977) and Sarukhán and

Piñero (in preparation), two patterns of life expectancy as a function of age emerge (Figure 8.3).

Although the general pattern of the curves is not too different, especially after the maximum e_x has been attained, two differences are worth pointing out. First, and probably due to the different lifespan of the species, a shift of the age with maximum life expectancy occurs, although the value of e_x is noticeably similar: around 55 and 65 years for *Astrocaryum* and *Euterpe*, respectively, despite the almost twofold difference in lifespan. Secondly, and more interesting, is the remarkable difference in the way the two species attain high life expectancies, after the very risky initial stages of life between birth and around 5–10 years of age. *Astrocaryum* climbs steadily to its maximum e_x at about 30–40 years, while life expectancy for individuals of *Euterpe* remains at a low 10 years for their first 35 years or so, to start sharply increasing after 45 years. This clearly reflects the relatively quick passage of individuals of *Astrocaryum* from the fewleaved 'seedling' stage, which lasts some 8 years, to the faster growing juvenile and immature forms whose crowns can be located up to some 3 m above the ground, which is virtually the situation of the mature, reproducing individual which keep slowly growing up to 6–7 m above the ground. On the other hand, individuals of *Euterpe*, a typical upper-storey palm, remain probably suppressed for some 35 years at heights of 3–4 m. Once the height at which suppression ceases is reached (at *c.* 9 m above the ground), either by sustained slow growth or by the effects of competition-liberating forest gaps, individuals of *Euterpe* improve six-fold their chances of future life. The first 8–10 years of *Astrocaryum* must then represent the suppressed stage which in *Euterpe* last four times longer.

The clearly different growth habits of the two palms under discussion invites one to consider whether their patterns of life expectancy through age would apply to other species, either those adapted to the relatively even microclimatic conditions within the forest or to those adapted to the variable macroclimatic conditions affecting the upper canopy of the forest. Accepting beforehand that there may be some inaccuracies, I used the estimated ages of individual of *Pentaclethra macroloba* as referred earlier in the paper, to calculate life expectancies for different 'ages' of the population. The results obtained fit strikingly well the expected pattern for an emergent species, similar to that of *Euterpe globosa*, especially in reference to the point of sharp increase in e_x values at about 50–60 years. At one point *Pentaclethra macroloba* departs from a smooth curve: at about 130 years the estimated value for e_x was 150 years while it should have been, according to the tendency of the curve, *c.* 240 (cf. Figure 8.3).

A further point that these curves suggest is that mortality processes, at least for populations of mature plants, are similar for widely different species with different growth habits. This corroborates the observation that the general pattern of mortality for populations of mature plants of many herbaceous species of temperate areas is very similar (Sarukhán & Harper 1973).

There are no other sets of data with which to verify the above suggestion, although a fair amount of observations on saplings and young individuals of emergent species seem to confirm these patterns of life expectancies.

A remark should be made here on the tendency to consider all seedlings and saplings growing in the lower layer of the forests as 'suppressed' by the taller trees. Such application may be warranted for saplings of 'light-demanding' species (Whitmore 1975), while it is not adequate for the 'shade-bearing' species. More research is needed on defining the influence of light regimes on the growth rates of species belonging to those two broad groups, and the consequences this has on plant survival.

REPRODUCTIVE SCHEDULES

Periodicity of flowering is a common event in tropical areas. Despite this, common event only recently ecologists and foresters have undertaken methodic and careful observations on the extent and frequency of the flowering process for a number of tropical tree species. Several general patterns have emerged, both in the times of the year when flowering occurs (Croat 1969; McClure 1966; Medway 1972; Smythe 1970; Whitmore 1975) and the temporal spacing of the flowering events (Burgess 1972; Frankie, Baker & Oppler 1974; Janzen 1967, 1974, 1976b, 1977; Smythe 1970).

Abundant evidence seems to suggest that, in a given year, flowering of many tropical lowland trees responds to water availability, although other factors may also be involved in the process (Alvim 1964; Burgess 1972; Fox 1972; Mori & Kallunki 1976; Poore 1968; Schulz 1960). However, the ultimate cause of these flowering frequencies and patterns is not yet clear. Floral biology and pollinator traits may act as selective forces which define flowering behaviour in tropical trees (Bawa & Opler 1975; Opler, Baker & Frankie 1975). Janzen (1977a) has proposed that predation has been 'the major driving force' in setting all the variations that exist in fruiting (and, by extension, in many instance, flowering) of tropical trees. Ever-increasing evidence of the importance of predators as reducers of plant numbers in the tropics lends support to this view.

These, or other hypotheses, may serve to explain the causality of some long-term behaviours, but they throw little light on how such trends are achieved, maintained or modified in the short term. It is indispensable, therefore, to determine accurately not only the flowering event in a species in a given year, but also how many and which individual did flower, how many of them will flower again at the next occasion, what is the real outcome of such initial reproductive expense (i.e. seeding efficiency), and what is the fate of seeds produced at the peak or the extremes of the flowering periods. All these questions fall directly in the realm of demography but such basic information is often not obtained, mostly on account of the shortness of the observations. I agree with Harper & White (1974), that 'it requires remarkable tenacity of purpose to collect data about the fecundity schedules of plants'.

Flowering and fruiting schedules within the life of tropical plants (whether herbaceous or woody) are, with very few exceptions, unknown. Ng (1966) found ages at first flowering for a large number of Dipterocarpaceae in Malaya to fall mostly between 20 and 30 years. Seeds of a number of these trees (among them species of *Dryobalanops, Shorea* and *Dipterocarpus*), showed very high viability. An astonishing record of flowering precocity, already observed by Kochumen (1961), was the flowering of 6-month-old seedlings of *Dipterocarpus oblongifolius*, but all flowers failed to set seed (Ng 1966). Bannister (1970) recorded plants of *Euterpe globosa* first flowering at *c.* 50 years, while the age at first fruiting for *Pentaclethra macroloba* (Hartshorn 1972) occurs at an approximate age of 180 years.

Observations on the number of flowers produced by plants are very scanty because of the many technical problems involved in estimating or counting flowers in normally dense inflorescences placed at many metres above the ground amongst the crown of the tree. Virtually nothing is known about general patterns of seeding efficiency of the flowers, i.e. what proportion of the flowers successfully turn into a fruit with viable seeds in it. However, it should be mentioned here that bearing seed may not be the sole purpose for which flowers are produced. Pollinator attraction and pollen production by functionally or morphologically male flowers are some alternative functions.

A few data on seeding efficiency of flowers of tropical trees exist. *Astrocaryum mexicanum*, for example, produced on average 28·4 female and 8 212·4 male flowers in an inflorescence (Sarukhán 1977; Sarukhán & Piñero, in preparation). Sixty per cent of the female flowers become fruits of which 94 per cent are viable. Apparently most of the 40 per cent of flowers that do not fruit is due to lack of space along the inflorescence axis for the fruits to develop. Consequently, an early pruning takes place before any more resources are allocated into developing fruits.

Opler *et al.* (1975) found, for various species of *Cordia* in Costa Rica, that seeding efficiencies varied between 15 and 30 per cent, a figure close to our observations on *C. elaeagnoides* in deciduous forest of the Pacific Coast in Mexico in which only between 15 and 20 per cent of the flowers bear seed, despite a fantastic and apparently 'useless' yearly flower crop.

Ataroff (1975), records flower crops for individuals of two savanna trees: *Byrsonima crassifolia* (from $3\cdot4 \times 10^3$ to 36×10^3 flowers/individual) and *Curatella americana* (from $21\cdot5 \times 10^3$ to 177×10^3 flowers/individual). Despite the ample range of flower crops, the proportion of flowers which produce a fruit is not so variable: for the same extremes 56 and 47 per cent for *Byrsonima crassifolia* and 74 and 56 per cent for *Curatella americana*. Heithaus, Opler & Baker (1974) found that *c.* 28 per cent hermaphroditic flowers of *Bauhinia pauletia* (or 12 per cent of the andromonoic flowers) set mature pods. Bawa (1973), in an extensive study of breeding systems of tropical trees, found wide variations in the number of open pollinated flowers that bore fruit; these vary from 0·1 per cent for *Enterolobium cyclocarpum*, to 1 per cent for species like *Guazuma tomentosa*, *Hirtella racemosa* and *Pterocarpus rohrii* and to more than 60 and 90 per cent for *Ardisia revoluta* and *Ochroma pyramidale*.

The cause of the inter or intrapopulation variability of flower-crop sizes and 'seeding efficiencies' is very poorly understood and their demographic consequences are simply unknown.

Perhaps the most complete observations to date on reproductive schedules, involving both flowering and fruiting, has been obtained for *Astrocaryum mexicanum* (Sarukhán 1977; Sarukhán & Piñero, in preparation). Because of the ease with which its reproductive parts can be reached, a very detailed accounting of the number of inflorescences, the number of male and female flowers and the number of fruits can be obtained for every reproducing adult in the population. Although only two complete reproductive cycles have so far been observed, interesting trends on reproductive schedules can be identified.

Averaging all permanent sites of observation, 45 per cent (with range 35–53 per cent) of all mature individuals (40–110 years old) fruited in 1975, a figure which was very closely repeated for the next year, only that 26 per cent (with range 7–42 per cent) of the adults which fruited in 1975 failed to flower in 1976. An average of 13 350 fruits ha^{-1} are produced in a year. Figure 8.4 shows the reproductive contribution of all age (size) classes of *A. mexicanum* for the year 1975, as well as the average number of fruits produced by individuals of different ages, starting from the youngest age at which flowering has so far been recorded (39 years), A fluctuating, but clearly ascending, curve of fecundity appears for this palm, without any indication of dwindling reproductive power due to age. In contrast, reproductive contribution per age or size category starts low and closely follows the same ascending trend of the average number of fruits per individual, but it clearly drops with age after *c.* 70 years, due to the progressively reduced number of individuals in the older categories.

Hartshorn's (1972) data for *Pentaclethra macroloba* (Table 8.2) shows an identical pattern of fecundity: a constant increase of number of seeds per tree from *c.* 185 years (80 seeds) to ages older than 350 years (357 seeds); percentage contribution starts also low (15 per cent at *c.* 185 years) reaches a maximum at between 240 and 300 years (a constant 30 per cent) and drops to 5 per cent at the maximum sizes (ages) recorded. The same pattern holds true for *Welfia georgii* (Vandermeer, unpublished data) where number of seeds/adult increase steadily from nearly 50 at the youngest reproductive stage to almost 900 at the oldest size class. Data on number of nuts/palm in *Cocos nucifera* (quoted by Harper & White 1974) also conforms to this trend.

Judging from the previous data, the generalization suggested by Harper & White (1974) that, following a juvenile phase, the fecundity schedule of trees increases to a maximum for some years and then declines steadily with age,

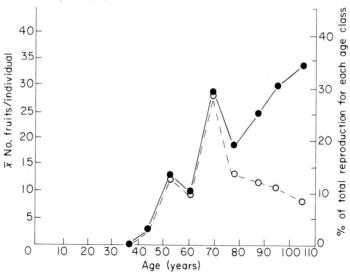

FIGURE 8.4. Reproductive schedules of individuals of *Astrocaryum mexicanum*. Data are averages of six 600 m² permanent plots (calculated from Sarukhán (1977)). Average number of fruits/individual (●——●), percentage of the total reproduction represented by each age category (○ – – – ○).

does not seem justified. This generalization was based on data which had been gathered mostly from very transformed, cultivated orchard plants (i.e. mango, Singh 1960), growing in rather artificial, man-made environments. The situation found in the naturally growing trees discussed suggests either that there are no senescence effects during the observed life-span of these species, or that plants are killed by accidents before senescence is reached.

I have referred so far to sexual reproduction in tropical plants. There is a large group of species with an exceedingly important life form: the ground-rooted climbing plants or lianas which in some cases depend on the production of ramets to increase their number of 'individuals'. Most descriptions and structural analyses of tropical vegetation characteristically ignore these plants, although they may prove to be, when better known in their biology and role in the forest, rather important components of the community. For example, in a tropical, semi-deciduous forest of the Pacific Coast of Mexico, where vines are not nearly as abundant and well developed as in the high evergreen forests, three species of vines (*Podopterus cordifolius, Entada polystachia* and *Bignonia unguis-cati*) represent 13·34 per cent of the total leaf litter produced by the 105 species of the forest in a year (Vizcaino & Sarukhán, in preparation).

The possession of well-developed, often massive, generally underground reserve organs (i.e. vines in the Dioscoreaceae, Fabaceae, Convolvulaceae) enables them to easily produce new shoots, often wide apart from each other, which serve as efficient canopy explorers.

TABLE 8.2. Fecundities found for individuals of different age (size) classes in species of tropical trees

Pinus caribaea (mean seeds/cone)	age (years)	10	20	30	40					(Hutchinson 1976)
		5	20	42	45					
Pentaclethra macroloba (seeds/adult)	d.b.h. (cm)	20<40	40<60	60<80	80<100	>100				(Hartshorn 1972)
		80	121	149	225	357				
Welfia gorgii (mean seeds/adult)	heights (m)	5·1–9·0	9·1–11·0	11·1–14·0	14·1–17·0	17·1–21·0				(Vandermeer, unpublished data)
		48·71	41·40	525·75	831·55	861·25				
Astrocaryum mexicanum (mean seeds/adult)	age (years)	43	52	60	68	78	86	94	104	(Sarukhán 1977; Sarukhán & Piñero, in preparation)
		3	13	10	28	19	25	30	34	

Through a study of two vine species, *Ipomoea phillomega* and *Marsdenia laxiflora*, Peñalosa (1975) shows these life forms to be beautiful subjects for analysis of an individual as an integrated population of plant parts, where a massive system of interconnected stems transports photosynthates from a canopy which is 40 m above the ground in the upper layer of the forest, to growing shoots that may be dozens of metres away in the forest ground. Peñalosa found that the totality of leaves in stolons of *Ipomoea phillomega* growing in low shade conditions are abscissed because of insect damage, 21 per cent of the damage to the leaves being total. In contrast, leaves of leafy stolons growing in deep shade drop mostly intact (only 12 per cent of the abscissed leaves showed insect damage). Half-life for the first type of leaves was found to be 6 days while it was nearly 24 days for the second type.

The fate of a stolon shoot is simply defined: it either twines quickly or its succulent tip will be grazed by herbivores. Stolon apices were found by Peñalosa to be considerably longer lived than twining apices. Survivorship of tips appears to be enhanced with ageing of the stolon or twiner. Loss of the tip by grazing represents an energetic loss that surpasses the amount actually sequestered by the herbivore, since it implies the mobilization of resources from very long distances to the decapitated shoot. Effects of tip grazing therefore may be quite limiting to survivorship of vines. Janzen (1971c), finds that up to 50 per cent of the tips of twiner shoots of *Dioclea megacarpa* are grazed by moth larvae.

No other sytems of vegetative 'reproduction' have been even superficially studied in the tropics, and climbers certainly merit more careful attention in their growth biology, as well as on their sexual reproduction. Peñalosa (1975) reports his being unable to find a single vine seedling in the many hectares of the study area; however, this group of plants flower and fruit profusely and their propagules have remarkable diversity in shapes, sizes and structures. Observations on seedlings of *D. megacarpa* (Janzen 1971c) suggest that, at least in some cases, comparative studies between vegetative and sexual means of recruitment can be carried out without technical difficulties in vines.

DYNAMICS OF
SEED NUMBERS

Seed population dynamics involves the voyage through several dramatically different situations of a highly specialized and characteristic plant structure, which, in addition, has to be somehow responsive to extremely complex sets of biotic and physical agents acting upon it, and which determine its successful transition to a potentially reproductive organism.

In its passage from being an organ attached to the parent plant to its incorporation into the soil reservoir of propagules, a seed almost unavoidably has contact with one or several animal species, contact that very often proves fatal for the seed. With the exception of the period in which the seeds remain dormant in the soil, determined mostly by physiological events, they are under the pressure of interactions with animals. The result of these interactions may be diametrically opposite: on one hand, seeds depend greatly on effective dispersers to be deposited in sites which insure high probabilities of successful establishment; on the other, predators directly, and ineffective dispersers indirectly, are the greatest causes of mortality for the plant population taken as a whole.

Seed predation, both pre- and post-dispersal, is very intense in the tropics and is often the result of an interaction between a highly specialized animal (frequently an insect) and a plant that has co-evolved chemical, physical and behavioural defences against the predator. A measure of such intensity may be the fact that seeds of most primary forest species germinate almost immediately after they land on the ground, even though their chances of long-term establishment are almost nil. Apparently, an individual has more probabilities of surviving as a competing seedling, than as a seed laying on the ground as a potential meal for a predator.

Many specific and general studies have made evident the highly complex structure of animal-plant interrelationships at the level of seeds (Bartholomew 1970; Janzen 1969, 1970, 1971a, 1976b; Kemp & Keith 1970; Krefting & Roe 1949; Newton 1967; Rosenzweig & Sterner 1970; Shaw 1968; Smith 1970).

Janzen's contributions (1971b,c, 1975a,b, 1976a, 1977a, in addition to those cited above) have set up what I consider as an important framework of carefully obtained factual information and theoretical ideas within which demographic studies can play an important role in exploring the finer (and therefore more basic) details of this plant–animal interrelationship.

The literature on seed predation reveals that the demographic conseqences of this interaction are enormous. Flowering and fruiting schedules of plants can be profoundly affected because of changes in predator pressure. *Hymenaea courbaril* strongly modifies not only the fequency with which its populations flower, but also its fruit morphology, depending on whether its predators are present or absent (Janzen 1975a). This obviously must affect the relative reproductive contribution of plants of different sizes or ages to the population, as well as rearranging allocation of energy expenditures.

Limited predator damage or parasitism in seeds of *Mucuna andreana* drastically lowers their seedling's fitness. For the same species, and for *Ateleia herbert-smithii*, Janzen (1977b,c) finds an ample variation in seed weights caused by differences in seed reserves; such variability is interpreted as adaptive in producing a more homogeneous seed shadow, because of the differential dispersal properties of the seeds. If this is the case, it would have important repercussions in establishing initial distribution patterns for the species as well as affecting seedling fitnesses.

Studies on seed predation of *Scheelea rostrata* (Janzen 1971b) suggest that predators may influence not only the flowering schedule by retarding age of first fruiting, but also the palm's sexuality by shifting genotypes to increase dioecism in the population. A single hectare of almost any type of tropical forest is full of interrelations of the kind discussed above.

Seed dispersal has been rather thoroughly reviewed and discussed at the superficial, 'macrodistribution' level (Ridley 1930; van der Pijl 1969) but this view throws little light on the subsequent predatory and demographic consequences of seed dispersal from a given parent plant, in a given forest situation and at a given time. Patterns of seed dispersal of tropical plants (either by wind or animal agents) are very scarce and allow only a very patchy image of the phenomenon. A common feature of all observations on dispersal distances by wind is that a majority of the propagules fall within a few dozen metres from the parent tree, even in cases where clear adaptations to wind dispersal are present; where such structures are poorly developed, maximum dispersal distances are sharply reduced (as in the case of several species of *Shorea* cited by Fox 1972 and Burgess 1970). Vertebrate seed dispersers produce much more heterogenous seed shadows and, frequently, the line that separates them from being predators of the seeds they disperse is very tenuous (Janzen 1971a; Vandermeer 1977b). They also show, in general, ample differences from fruiting cycle to fruiting cycle in the intensity with which they interact with the seeds, either as dispersers or predators. This obviously should have important impact on the demography of the plants concerned. So far, most data on seed dispersal and predation by vertebrates 'do little more than point to annoying complexities introduced into the life history of a plant due to its having mammal dispersed seeds' (Vandermeer 1977b,c).

Most seeds of tropical stable forests opt for germination as a means to avoid predation, so build-up of seed populations in the soil is not a good reflection of 'predator escapes'. Judging by the available data and assuming that these are representative of the forests sampled, soil seed banks constitute a relatively species-poor component of the ecosystem. Keay (1960) finds in a sampling area of between 0·4 and 0·8 m^2, 43 species (15 trees, 4 shrubs, 7 climbers, 13 non-climbing herbs and 3 'doubtful plants'); the average number of seeds was 233 m^{-2}. In *Dacryodes excelsa* montane forests in Puerto Rico, Bell (1970) found, in a 1·5 m^2 sampling area, only 13 species with a total number of seeds which varied from 152 to 424 m^{-2}. Guevara & Gómez-Pompa (1972) found in a sample of 0·5 m^2, taken in 2-month-old secondary vegetation, 23 species and 800–2 700 seeds m^{-2}; in a 5-year-old secondary stand 19 species and 2 000–4 000 seeds m^{-2}; and in primary rain forest 13–26 species and 175–900 seeds m^{-2}. In contrast with what happens in temperate areas, studies of seed banks in the soil of agroecosystems are even more poorly studied than those of natural systems. Kellman's (1974) study of agricultural soils of Belize shows in a relatively small sampling area (*c.* 230 cm^2) a much greater diversity of seed flora: 54 species and a total of 6 497 seeds m^{-2}. One other study (Miége & Tchoumé, cited by Kellman 1974) in agricultural soils in Senegal, obtained very similar results.

However interesting these studies are, they tell us little on how such banks

TABLE 8.3. Probabilities that a fruit of *Astrocaryum mexicanum* borne in the infructescence axis on the palm, will be ready to germinate on the ground (Sarukhán 1977)

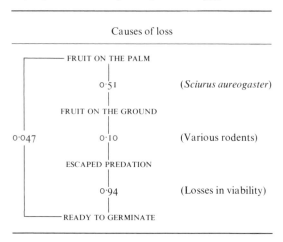

Causes of loss

FRUIT ON THE PALM

0·51   (*Sciurus aureogaster*)

FRUIT ON THE GROUND

0·047   0·10   (Various rodents)

ESCAPED PREDATION

0·94   (Losses in viability)

READY TO GERMINATE

are built and maintained and the dynamics of the seed populations that may compose them: besides, studies on the viability of seeds under soil conditions (Juliano 1940; Mensbruge 1966), do not help much either when they constitute only 'pecking' samples of the soil in a forest. Methodic and careful observations are needed to support the generalized idea of the great longevity of seeds of species typical of secondary vegetation, deduced only from the large numbers of seeds of these species found in the soil. No study of soil seed banks has been able to distinguish seeds already *in situ* from the continuous reseeding by dispersing vertebrates. Nor is there a study involving periodic, continuous sampling of the seed bank in the soil for at least a few years in permanent sites, to record changes in composition and number in the seed banks.

Studies on entire seed population dynamics of tropical plants are few and none of them is as yet very intergrated. In *Astrocaryum mexicanum* (Sarukhán 1977), predation of the one-seeded fruits can take place both when they are still attached to the infructescence on the palm (mostly by the arboreal squirrel *Sciurus aureogaster*) and on the ground, after they have fallen (by several species of rodents) (Table 8.3). Our observations show that predator effects are highly variable and, although fruits have in general greater chances of escaping predation in sites of higher than of lower palm densities, the data obtained so far does not permit any conclusive answer to whether palm or fruit densities as such affect predation rates.

On the whole, chances that a fruit of *Astrocaryum mexicanum* in the axis of the infructescence will reach the point of germination are 0·047, mostly due to a 95 per cent predation rate. Similar predation rates have been found for fruits of other forest palms like *Scheelea rostrata* (Janzen 1971b), where bruchids and rodents kill 90·6 per cent of all fruits, an extra 7·5 per cent being unviable. Removal rates by different vertebrates of fruits of *Welfia georgii* may vary in a complex way from 7 to over 80 per cent (Vandermeer 1977b). Bruchids can take from 0 to 100 per cent of the seed crop of *Dioclea megacarpa*, in addition to an extra 7 to 42 per cent mortality caused by squirrels (Janzen 1971c). Hartshorn (1972) finds that seeds of *Pentaclethra macroloba* are virtually predator free. Parrots (*Pionus senilis*) and squirrels (*Sciurus variegatoides*) account for 6·5 of the total 10·4 per cent; the rest is mortality caused by bruchids.

Synott (1975) finds that rodents cause an average 43 per cent mortality of seeds of *Entandrophragma utile* in Uganda. This average masks an enormous variability of seed predation that ranges from 90 to 95 per cent in areas of stable mixed forest to 3–10 per cent in secondary vegetation or forest-edge communities.

In a study of the dynamics of seed population of *Cordia elaeagnoides* in the soil (Guevara 1977), two series of experiments in contrasting soil depths were established in tropical deciduous forest. For both series, the seed bank was reduced to 3 per cent at the second month after the experimental introduction of seed into the soil, and was under 1 per cent after the fifth month. All losses from the seed bank were caused by predation, since maximum germination recorded *in situ* never exceeded 1·5 per cent. Two small rodents, *Liomys pictus pictus* and *Oryzomys palustris*, but particularly the first, are responsible for the losses to the seed bank of *Cordia elaeagnoides*. It must be noted that only 15–18 per cent of the fruits sown contained seed (it is impossible to differentiate between viable and unviable fruits by external inspection); despite this, the rodents did not show any preference in selecting their food between both types of fruits. It is not known yet if such low viability is a constant feature in the reproduction of *C. elaeagnoides*, or whether it is cheaper to produce the unviable nutlets so they can be used as a 'camouflage' strategy by the plants to satiate rodents in order to save, at least, some costlier viable fruits.

It is fair to think that a point has been reached in the general field of seed predation where significant progress in exploring the basic nature and meaning of the phenomenon will depend on detailed demographic approaches.

DYNAMICS OF SEEDLING NUMBERS

The features that increase seed survival are not necessarily beneficial for seedling survival. Often there may be a lack of correlation or even contradiction between the adaptive traits of seeds and seedlings. Much of the same level of complexity in the requirements for establishment and growth of seedlings found in species of temperate areas occurs in tropical plants. Numerous studies on seed germination requirements, seedling establishment and growth, adaptation to light, moisture, temperature regimes, etc., have been carried out, especially for economically important (timber) species (Bieleski 1959;

Chapter 8

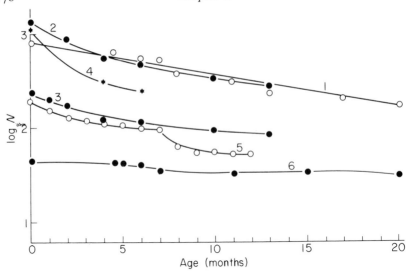

FIGURE 8.5. Survivorship curves for cohorts of seedlings of tropical trees: 1. *Nectandra ambigens*, cohort 'older than 1975'; 2. *N. ambigens*, cohort of 1976, site 1; 3. *N. ambigens*, cohort of 1976, site 2; 4. *Entandrophragma utile* (Synnott 1975); 5. *Dipterodendron costarricense* (Fournier & Salas 1967); 6. *Astrocaryum mexicanum*, data average of six 200 m² permanent plots for seedlings of various ages.

Coombe 1960; Coombe & Hadfield 1962; Dawkins 1956; Goodall 1949; Landon 1955; Nicholson 1960; Njoku 1964; Synnott 1975): it is beyond the aims of this chapter to explore this subject in detail. I will restrict the discussion on numeric changes in seedling populations to the biotic factors that may be involved in such changes and their demographic implications.

Although the observation and study of seedling populations presents few problems with age determination, since differentiation among plants in their first months of age is minimal, reliable information on the fate of seedlings of tropical trees is very meagre. Even studies aimed at obtaining demographic data on a species not always incorporate seedling information in a complete and accurate fashion. Forestry records of regeneration are affected by the extremely lax use of the word 'seedling' (any plant not reaching the nose of the surveyor is often lumped with true, new-germinated seedlings) and the absence of ecological insight which makes the careful recording of marked seedlings of different cohorts in permanent plots, for long periods, a veritable oddity in tropical forestry studies, despite the very great importance that is given to species regeneration in silvicultural management (Philip 1968; Wyatt-Smith 1955).

Survivorship curves for a few cohorts of seedlings of several species are shown in Figures 5a, b. All survivorships curves for *Nectandra ambigens* are virtually replicated for the two different populations of the same 1976 cohort, but differ from the 'older than 1976' mixture of cohorts, already present when the study began, which is wholly exponential. It is important to remark that, although the forest floor where *N. ambigens* grows in Southern Veracruz is usually carpeted with seedlings (specially after seed fall), there are no intermediate sizes between these and the fully mature, emergent trees and no records of human disturbance for the study forests exist. This species is considered as 'light-demanding'.

'Half-lives' for the 'older than 1976' group of seedlings is around 7·5 months, a figure which is not too different from that found by Smith (1970) of 6·5 months for 'seedlings' (of all species) of a Puerto Rican forest. Once exponentiality is attained by the curves of the 1976 cohorts of *N. ambigens*, their survivorship does not greatly differ from the 'composite' exponential curve, although 'half-lives' increase to 8·5 and 16·5 months for each population recorded.

Dipterodendron costarricense (Fournier & Salas 1967) seedlings follow closely the same mortality patterns of *Nectandra ambigens* up to the seventh month of life after which a sudden increase in mortality occurs. This is

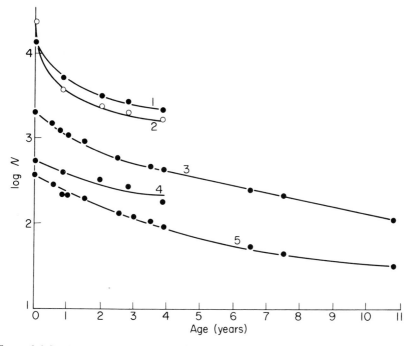

FIGURE 8.6. Survivorship curves for cohorts of seedlings of Tropical trees. 1. *Parashorea tomentella* (Liew & Wong 1973); 2. *Shorea gibbosa* (Liew & Wong 1973); 3. *Shorea parvifolia* (Wyatt-Smith 1958); 4. *Shorea ovalis* (Liew & Wong 1973); 5. *Koompasia malaccensis* (Wyatt-Smith 1958).

interpreted by the authors as the result of the onset of the dry season in the study area, which was in a 12-year-old secondary forest, probably very exposed to the wider fluctuations of the macroclimate. Such periodic increases of mortality risks have also been observed, although not so drastically as in the former example, for seedling cohorts of *Astrocaryon mexicanum* originated before 1975 (Sarukhán & Piñero, in preparation) (cf. Figure 8.5); although the average curve presented smooths out the trend, the greater mortality rates correspond with time of germination of the new cohorts (November–February). The short dry season in the area where *A. mexicanum* grows occurs later in the year (April–May), a time of virtually 100 per cent survival of seedlings. We have no evidence up to now that accidental causes of mortality (i.e. falling branches or trees) or any microclimatic factor may account for the simultaneous increase of seedling mortality of *A. mexicanum* in the six populations studied.

A similar, very smooth, Deevey type III curve is obtained when plotting the longer observations on seedlings of several Dipterocarpaceae (Figure 8.6), with the exceptions of *Parashorea tomentella* and *Shorea gibbosa* (Liew & Wong 1973), which show heavier mortalities at early stages of seedling life, as is also in part the case for *Entandrophragma utile* (Synnott 1975). Half life of seedlings of *Shorea parvifolia*, after the exponential rate is attained, is *c.* 3·5 years. No clear, generalized trends can be identified among the survivorship curves on Figures 8.5 and 8.6 and mortality rates of Table 8.4 apart from a reduction of death rates with age. More and longer observations on seedling cohorts are needed.

All seedlings on the forest floor may be said to be under a very delicate

TABLE 8.4. Comparison of percentage mortality rates for the first 2 years of age of seedling cohorts of tropical trees

Species	Mortality (%) Age of cohorts (months)				Source
	6	12	18	24	
Nectandra ambigens					
(1976, cohort a)	63·9	39·6	—	—	
(1976, cohort b)	49·8	25·8	—	—	
Dipterodendron costarricense	56·5	45·9	—	—	Fournier & Salas (1967)
Entandrophragma utile	76·8	—	—	—	Synnott (1975)
Shorea parvifolia	35·5	14·5	22·6	20·4	Wyatt-Smith (1958)
Koompasia malaccensis	27·4	21·9	0	22·2	Wyatt-Smith (1958)
Parashorea tomentella	44·4	29·2	24·2	16·9	Liew & Wong (1973)
Shorea gibbosa	78·6	29·1	26·0	20·7	Liew & Wong (1973)
S. ovalis	13·9	14·8	16·6	13·3	Liew & Wong (1973)

metabolic balance, some (the light demanders) more than others (the shade tolerant). Under these conditions, many factors tip such balance down the negative slope, causing seedling death. Accidental causes of mortality may, in some cases, be important. Hartshorn (1975, 1977) and Vandermeer (1977a) report heavy mortalities of *Pentaclethra macroloba* and *Welfia georgii* seedlings, respectively, due to falling palm leaves and tree limbs. Trampling animals (buffaloes and elephants, for example) may also cause considerable seedling losses (Fox 1972; Synnott 1975).

Many of the post-dispersal seed predation interactions, especially by vertebrates, apply to seedlings. Post-dispersal seed predators often become seedling predators at low levels of food availability. I do not know of any study exploring whether many structural characteristics of seeds are at least equally adaptive at increasing seedling fitness against predators. For example, the extremely rough, bony endocarp of *Calatola laevigata*, covers the epicotyl of the germinating seedling, protecting it effectively against grazers until the seedling is some 20 cm tall and has enough lateral buds and well-developed hypocotyl to stand grazing damages. The huge woody drupes of *Andira inermis* serve the same purpose. Many species in the Anacardiaceae, Euphorbiaceae, Hippocrateaceae, Lauraceae, Sapotaceae, etc. of the neotropics, present this type of 'armour' protection for their young seedlings. Other species may combine such protection with extremely fast rates of growth, once germination is initiated, developing large, thick hypocotyls which will help recovery of seedlings whose tips may have been grazed. Some examples of this are *Omphalea oleifera, Poulsenia armata* and *Calophyllum brasiliense* in the neotropics. Whitmore (1974) has found the same behaviour for South-east Asian species of *Calophyllum.*

Most of the knowledge on seedling mortality due to predation is anecdotal; experimental studies of the type carried on *Mucuna andreana* (Janzen 1976a) are needed. Mimicked partial bruchid damage on seeds of this vine reduced the seedling's ability to withstand herbivory, simulated by controlled clippings of seedling tips.

Mortality of seedlings is mostly caused by predation, competition and accidental events. Such mortality factors acting on an existing matrix of seeds in the soil, originate the first recognizable basis of species patterns in the forest.

POPULATIONS OF MATURE PLANTS

All trees in the tropics have a finite life span. The 'calm serenity of the tropical forest' belies the intense mortality rates of mature trees most of which never reach the dimensions of the 'forest giants'. Between seedling establishment and the attainment of maturity, many events shape the species-specific way in which proportions of differents sizes (ages) of individuals compose a stable population.

Tropical forestry literature abounds in examples of stand tables or size distribution for a vast number of timber species, and different patterns of size distributions have been described (Schulz 1960; Whitmore 1974; Wong 1973). Stand tables have been limited to the analysis of girth classes because of the difficulty of relying on a structural 'age recorder' in most tropical trees.

Despite their dimensions, tropical trees do not attain records of longevity, compared to temperate trees. Radiocarbon datings of 800 ± 70 years for *Shorea curtissii* (Whitmore 1975) and $1\,000 \pm 100$ for *Adansonia digitata* (Swart 1963) are probably the oldest ages registered for tropical trees.

Age and size are often poorly correlated. Nicholson (1965) records 5·6–6·5-fold differences in age at maximum size attained by emergent and under-canopy trees, respectively, in forests of Sabah.

The dilemma of using age or size as pivotal characters in plants has been stressed by Harper & White (1974) who affirm that age is a poor predictor of reproductive activity (but see chapter 2). This is particularly true for herbs which may alter plastically their physiological status in a dramatic way. It is also true for very dense natural or planted monospecific stands of trees. But in other cases (e.g. *Astrocaryum mexicanum* and *Euterpe globosa,* referred to earlier in the paper) age may be a useful indicator of a plant's physiological status and reproductive performance. An important point, then, is to find out whether, for a particular species, age can be adequately used as a pivotal character. The inadequacy of its use should not in any way preclude the possibility of carrying out demographic studies, but it certainly constitutes an important advantage in determining age composition of a population since this allows establishment of many correlations to environmental and biotic events that otherwise could not be made.

Competition has influenced most of the silvicultural management techniques of tropical forests. Despite this, it is a particularly ill-known phenomenon, mostly because the study of competition among trees in a very complex community is fraught with all kind of methodological difficulties. Virtually all that is known about competition among forest trees concerns exclusively its effects on growth rates of size categories (Dawkins 1956). The nature of the phenomenon, its influence on reproductive behaviour and hence its demographic consequences, are wide-open fields for research in tropical systems.

Plant-animal relationships, mostly through grazing activities, may also affect the behaviour (specially reproduction) of mature trees (Simmonds 1951).

COMPETITION, PREDATION AND PATTERNS OF SPECIES DISTRIBUTION

The patchy distribution of mature trees of different species is a well-known feature of most tropical forests and has drawn the attention of many ecologists who have approached the problem from very broad analyses (Aubréville 1938;

van Steenis 1958) to more rigorous quantitative studies (Austin & Greig-Smith 1968; Webb *et al.* 1967; Webb, Tracey & Williams 1972; Williams *et al.* 1969). So far, these studies have mostly described static patterns, not having advanced much into the description, let alone the explanation, of the dynamic changes that occur in such patterns in space and time.

As mentioned a little earlier, the foundations of distribution patterns of mature individuals reside first on the distribution of seeds and later of seedlings. Therefore, understanding the factors which cause these basic patterns should be essential to understanding the dynamics which generate large-scale patterns in the forest. Some of the factors which affect seed distribution have been discussed already. How mortality factors (competition, predation, accidents) affect seedling distribution in the forest is poorly known.

Competition has long been considered an important factor in determining establishment and growth of forest seedlings (Kozlowski 1949; Shirley 1945), acting mostly through the availability of soil moisture and light, although it has been occasionally disputed (Fedorov 1966). Results from trenching and weeding experiments in tropical forest suggest (although sometimes nebulously) that competition mechanisms largely await to be experimentally explored.

Density-dependence mechanisms influencing establishment of *Welfia georgii* seedlings have been found (Vandermeer 1977a). Seedlings growing within 3 m of an adult palm have a lower probability of survival (0·42) than those further away (0·70). No relation of neighbour distance and seedling growth rates were found. It is not clear, however, whether competition, predation or accidental causes determine such survivorship patterns.

Connell (1971) found also that young saplings of trees in rain forests in Queensland, having a conspecific individual as nearest neighbour, had a higher mortality rate than those whose nearest neighbour was not conspecific. This pattern did not hold for older saplings. Seedings planted under a conspecific adult tree had a greater mortality than when planted under an adult of a different species. Herbivore grazing played an important role in determining mortality. This 'phytometer' approach seems very promising in devising experiments under natural (although complex) conditions to explore the nature of interference among seedlings and between seedlings and older individuals.

Preliminary analyses of pattern distribution in *Astrocaryum mexicanum* have shown that, in general, the very patchy pattern characteristic of seeds and seedlings evolves with the age of the individuals to an almost random distribution of the oldest palms. This conforms with Connell's (1971) observations that pattern in tropical forests becomes more diverse as older individuals are considered and as the community itself ages. We have also found that seedlings and juvenile plants, particularly, are subjected to a very rigid distribution pattern that seems to be density regulated. Seedlings and juveniles in sites of high *A. mexicanum* density (site A, 427 palms 600 m^{-2}) are restricted in their position in the forest floor: 95 per cent of seedlings and juveniles occur within a band of 40 cm at both sides of the vertical projection of the crown of adult palms; this occurrence is very significantly higher than that expected by pure chance. A similar situation is encountered when analysing the total population at this site: 95 per cent of all palms are within 75 cm of the vertical projection of the crown of another palm. This suggests that a subordination effect is taking place in dense *A. mexicanum* populations. For the low density sites (site C, 115 individuals 600 m^{-2}), seedlings and juveniles occur near the projection of the crown of an adult palm not more often than chance alone would determine. What causes these distributions is not clear to us. Competition for light (both qualitatively *and* quantitatively) is an aspect under investigation, but also predators may be responsible for the rigid location of seedlings and juveniles.

Herbivore activities (mostly through grazing) can have a strong influence on competitive interactions among plants and, therefore, this creates a very complex feedback mechanism between competition and predation, besides profoundly affecting species diversity in a community (Harper 1969).

Analysis of species pattern at this level of individual populations becomes a fascinating study subject which may throw much light on the nature of biotic relations between organisms.

POPULATION GROWTH AND INDIVIDUAL GROWTH: ENERGY ALLOCATION

When discussing strategies of energy allocation by plants, one can theoretically move from dealing with increments of components of a population (individuals), to the increments of components of an individual (structural units). Both have effects over each other and the rates of increase of one set of components sooner or later has effects on the rate of increase of the other set of components.

Harper & White (1974) have discussed the relevance of adaptations of structural units to the capacity of an individual to contribute to its population's growth. Studies on the architecture of trees, mostly with reference to their branching patterns (Hallé & Oldeman 1970; Horn 1971; Tomlinson & Gill 1973) have set up good structural and morphological frames of reference that should promote experimental studies on the ecological and adaptive significance of such energy-capturing systems.

Plant morphology cannot be defined in space alone; there is a very important time component which makes such pattern of growth and development a recurring cycle, repeated over the generations and subjected to selective pressures. These patterns determine to a large extent the allocation by a plant of its available energy.

It is of prime demographic interest to know how individuals of different ages and growing under different environmental conditions allocate their resources to different ends, and whether these decisions constitute successful strategies which maximise their overall fitness. An ample and well-known theory of resource allocation exists which has been adequately discussed in the context of plant populations (Gadgil & Solbrig 1972). The very few direct observations on this subject for plant populations have been carried out only with temperate herbaceous species.

Apart from a very gross estimation of energy allocation in plants of *Euterpe globosa* (Van Valen 1975) and an isolated observation on dry matter partitioning in an individual of *Corypha elata* (Tomlinson & Soderholm 1975), I know of no other study on the subject for tropical plants. Detailed studies on energy allocation in palms of *Astrocaryum mexicanum* have been carried out and a preliminary discussion on standing biomass and its distribution in the plant for individuals of all ages has been published (Sarukhán 1977). Further studies (Piñero, Sarukhán & Alberdi in preparation) have allowed us to assess the total energy devoted to each plant part by an individual during its whole life. A palm *c.* 120 years old has spent the following percentages of the *total* dry matter processed up to that age: for roots and stump 5·5 per cent, for the trunk 13·5 per cent, for leaves 46 per cent, giving a total of 65% per cent of biomass devoted to vegetative organs. For reproductive ends, the plant would have devoted 30 per cent of its total biomass to produce fruits and an extra 5 per cent for reproductive accessories (male flowers, spathes, inflorescence axis). Coincidental with the initiation of reproduction, plants at the first reproducing age start increasing noticeably their energy allocation to leaves by raising the leaf production rate. In real figures, the above data would mean that a palm *c.* 120 years old would have produced, in terms of kilogrammes of dry matter, approximately 13 kg of fruits, 3 kg of reproductive accessories, 5 kg of roots and stump, 7 kg of trunk and nearly 22 kg of leaves, giving a total dry weight of about 50 kg.

Calculating the yearly dry matter production by palms of all ages, expenditure to all plant parts in the yearly energy budget can be estimated. For example, an individual *c.* 120 years old spends 5, 11 and 47 per cent of its yearly budget for roots and stump, trunk and leaves, respectively: for reproductive structures it uses 33 and 4 per cent for fruits and reproductive accessories, respectively. These are very similar figures for the proportions of the total expenditure for an individual of this age presented above. An individual belonging to the category which has the largest reproductive contribution to the population (about 75 years), will have a very similar yearly budget as the oldest palm recorded, perhaps with *c.* 2–3 per cent less dry matter devoted to reproduction. This enforces the view derived from the reproductive schedules (cf. Figure 8.4), in which no senescence effects can be detected on the reproductive (or indeed any other) activity of very old palms.

These results depart considerably from the general patterns proposed by Harper & Ogden (1970) and Ogden (1968), on proportional expenditure by annual and perennial species. The situation found for *A. mexicanum* may be prevalent for many other perennial iteroparous plants where a clear distinction is made between the accumulated and shed growths in previous years and the current year's growth. No such analysis exist to compare the data of *A. mexicanum*. If this situation is verified for other perennials, the actual ideas about 'typical' energy expenditures of '*r*' and '*K*' plant species should be reconsidered.

CONCLUSION

Throughout this chapter, I have tried to emphasize the importance of demographic studies to the understanding of tropical population biology. I have maintained the view (Sarukhán 1977) that studies on the demographic events of a single species may help more than any other approach in exploring the intricate interactions and dynamics of a community, perhaps more significantly so in the complex tropical communities.

In evaluating information on different aspects of tropical plant demography, I have pointed out several specific problems, both on the phenomenon itself as on the information available or the lack of it. Perhaps the most important overall demographic problem in the tropics may be said to be the scarcity of demographic studies. This may also be viewed optimistically as the rise of a branch of plant population ecology which may, from the beginning, optimize its efforts to attain a better understanding of how tropical plants and their communities operate.

Demographic studies are time-consuming and effort-demanding pursuits, unattractive to 'fast-result' seekers. When tackled, a number of 'procedure norms' should be followed so results are rewarding to the researcher and useful for building a firm body of knowledge on the subject. Some basic procedural norms are:

1   Selection of the species to be studied must be based on solid knowledge of the community where it occurs, its structure and the apparent role played by the species in the community. The more important it is (physiognomically and ecologically) the more attractive does it become as a study subject. Plant demography in the tropics (and elsewhere) should be first built with the aid of demographically 'co-operative' species.

2   Studies should be inevitably made on permanent plots containing representative samples of populations of the species. There are no general criteria to define sample sizes for demographic studies, but pattern analyses may help in determining approximately such sample areas for a species. As a very coarse 'rule of thumb', population numbers in each site should be counted in hundreds, occasionally even in thousands. 'Replications' of each condition studied are always advantageous.

3   Such permanent sites must guarantee the realization of long-term (no less than 10 years) and very frequent observations (from almost every day to 3–4

per year, depending, for example, if one is observing newly born seedling cohorts or is checking leaf production rates). The guarantee must extend to the preservation of the conditions under which the study is meant to be carried out. In this respect, field stations or biological reserves play a very important role and, if properly managed, they should always be used for these studies.

4   Various conditions under which the selected species grow (e.g. density ranges, physical environmental gradients, community gradients, etc.) should be studied to enhance the information obtained and the representativeness of the data.

5   Detailed monitoring of every plant of the species selected should be possible, so almost any change occurring in the population can be reproduced 'on the desk'. The other components of the community should be known, at least in what refers to their diversity, abundance, size and relative position to the individuals of the selected species.

6   All aspects of behaviour (phenological and reproductive) relevant to the demography of the species should be recorded. Seedlings particularly require detailed and prolonged observation of survivorship and growth.

7   It can be greatly advantageous to the analysis of demographic data if age can be directly or indirectly (reliably) used. This is particularly difficult with tropical trees, but I would not be surprised if careful studies prove this difficulty to be more apparent than real in many cases. Xylem rings are not the only recording device in trees.

8   Lastly, experimental approaches should be used whenever they can be applied and enough knowledge of the phenomenon warrants their adequate use. Seed and seedling stages are particularly adequate for rigorous experimental analysis.

Besides the scarcity of good studies, tropical plant demography suffers from acute imbalance. No studies on biological forms other than trees exist nor ecosystems other than evergreen forests have been considered. I do not think it necessary to insist in how much awaits to be explored demographically in the tropics.

A major 'tropical system' that is particularly interesting (conceptually and methodologically) for demographic studies is the enormous variety of secondary communities derived from almost every stable forest in the tropics. These communities pose a deluge of intriguing questions and very little of their dynamics has been studied. Additionally, these systems are becoming (and soon will be) the most widespread plant communities in the world tropics and should be far better known than they are today. As a result of the overwhelming rate of human disturbance, well preserved, primary tropical forests are shrinking steadily. However, this is not the only effect. Many stable forests are being strongly modified in their faunistic endowments and this surely has a profound impact on the dynamics of the components of the forest. This is a point which should be borne in mind, particularly when one deals with seeds, seedlings and their dispersers-predators.

The latter consideration means that in many areas a race against time must be held if primary forests are to be studied before they become too modified by peripheral disturbance. But, alas, racing against time is the great handicap of demographic studies!

References

Abbot R.J. 1976. Variation within common groundsel, *Senecio vulgaris* L. I. Genetic response to spatial variation of the environment. II. Local differences within cliff populations of Puffin Island. *New Phytol.* **76**:53–72.

Abrahamson W.G. 1973. *Resource Allocation in Plant Populations of Different Habitats.* Ph.D. Thesis. Harvard Univ. (Cambridge).

Abrahamson W.G. 1975a. Reproductive strategies in dewberries. *Ecology* **56**:721–726.

Abrahamson W.G. 1975b. Reproduction of *Rubus hispidus* L. in different habitats. *Amer. Midl. Nat.* **93**:471–478.

Abrahamson W.G. 1979. Patterns of resource allocation in wildflower populations of field and woods. *Amer. J. Bot.* **66**: 71–79.

Abrahamson W.G. and M.Gadgil. 1973. Growth form and reproductive effort in goldenrods (*Solidago*, Compositae). *Amer. Nat.* **107**:651–661.

Adams C.D. and H.G.Baker, 1962. Weeds of cultivation and grazing lands. In Willis J.B. (ed.), *Agriculture and Land Use in Ghana*, pp. 402–415. London, Oxford University Press.

Adams M.L. and O.L.Loucks, 1971. Summer air temperatures as a factor affecting net photosynthesis and distribution of eastern hemlock (*Tsuga canadensis* L. (Carriere)) in southwestern Wisconsin. *Amer. Midl. Nat.* **85**:1–10.

Ahlgren C.E. 1974. Effects of fires on temperate forests: North Central United States. In Kozlowski T.T. and C.E.Ahlgren (eds.), *Fire and Ecosystems*, pp. 195–223. New York, Academic Press.

Allard R.W. 1960. *Principles of Plant Breeding.* New York, Wiley.

Allard R.W. 1961. Relationship between genetic diversity and consistency of performance in different environments. *Crop Sci.* **1**: 127–133.

Allard R.W. 1965. Genetic systems associated with colonizing ability in predominantly self-pollinated species. In Baker H.G. and G.L.Stebbins (eds.), *Genetics of colonizing species*, pp. 49–75. New York, Academic Press.

Allard R.W. 1975. The mating system and microevolution. *Genetics* **79 (Suppl.):**115–126.

Allard R.W. and J. Adams, 1969. Population studies in predominantly self-pollinated species. XII. Intergenotypic competition and population structure in barley and wheat. *Am. Nat.* **103**:621–645.

Allard R.W., S.K.Jain and P.L.Workman, 1968. The genetics of inbreeding populations. *Adv. Genet.* **14**:55–131.

Allard R.W. and P.L.Workman, 1963. Population studies in predominantly self-pollinated species. IV. Seasonal fluctuations in estimated values of genetic parameters in lima beans populations. *Evolution* **17**: 470–480.

Allen D.C. 1976. Methods for determining the number of leaf-clusters on sugar maple. *For. Sci.* **22**:412–416.

Allen E.B. and R.T.T.Forman, 1976. Plant species removals and old-field community structure and stability. *Ecology* **57**:1233–1243.

Allen M. 1977. *Darwin and his Flowers: the Key to Natural Selection.* London, Faber & Faber.

Alvim P.de T. 1964. Tree growth periodicity in tropical climates. In Zimmerman M.H. (ed.), *Formation of Wood in Forest Trees*, pp. 479–495, New York, Academic Press.

Amen R.D. 1974. Perspectives on the condition of seed dormancy. *Trans. Amer. Micro. Soc.* **93**:593–596.

Anderson R.C. 1965. Light and precipitation in relation to pine understory development. M.S. Thesis. Univ. Wisconsin (Madison).

Anderson R.C. and O.L.Loucks, 1973. Aspects of the biology of *Trientalis borealis* Raf. *Ecology* **54**:798–808.

Ando T. 1962. Growth analysis on the natural stands of Japanese red pine (*Pinus densiflora* Sieb. et Zucc.). II. Analysis of stand density and growth. *Gov. For. Expt. Stn. Tokyo Bull.* **147**.

Andrews H.N. 1963. Early seed plants. *Science* **142**:925–931.

Anon. 1970. Leaf angle of corn plants may affect yield. *Crops & Soils* **3(3)**:24.

Antonovics J. 1968. Evolution in closely adjacent plant populations. *Heredity* **23**:219–38.

Antonovics J. 1972. Population dynamics of the grass *Anthoxanthum odoratum* on a zinc mine. *J. Ecol.* **60**:351–366.

Antonovics J. 1976. The nature of limits to natural selection. *Ann. Mo. Bot. Gard.* **631**:224–247.

Apirion D. and D.Zohary, 1961. Chlorophyll lethal in natural populations of the orchard grass (*Dactylis glomerata* L.). A case of balance polymorphism in plants. *Genetics* **46**:393–399.

Appleby A.P., P.D.Olson and D.R.Colbert, 1976. Winter wheat yield reduction from interference by Italian ryegrass. *Agron. J.* **68**:463–466.

Armstrong R.A. and M.E.Gilpin, 1977. Evolution in a time-varying environment. *Science* **195**:591–592.

Arnott R.A. 1969. The effect of seed weight and depth of sowing on the emergence and early seedling growth of perennial ryegrass. *J. Brit. Grassl. Soc.* **24**:104–110.

Arroyo M.T.Kalin de, 1976. Geitonogamy in animal pollinated tropical angiosperms. *Taxon* **25**:543–548.

Ashton P.S. 1969. Speciation among tropical forest trees. *Biol. J. Linn. Soc.* **1**:155–196.

Aspinall D. 1960. An analysis of competition between barley and white persicaria. II. Factors determining the course of competition. *Ann. Appl. Biol.* **48**:637–654.

Ataroff M. 1975. *Estudios ecológicos poblacionales en dos especies de árboles de las sabanas de los llanos. Tesis licenciatura, Facultad de Ciencias, U. de los Andes, Venezuela.*

Aubréville A. 1938. La forêt coloniale: les forêts de l'Afrique occidentale française. *Annals Acad. Sci. colon.* **9**:1–245.

Auclair A.N. and G. Cottam, 1971. Dynamics of black cherry (*Prunus serotina* Ehr.) in southern Wisconsin oak forests. *Ecol. Mon.* **41**:153–177.

Austin M.P. and P.Greig-Smith, 1968. The application of quantitative methods to vegetation survey. II. Some methodological problems of data from rain forest. *J. Ecol.* **56**:827–844.

Bailey R.J., H.Rees and L.M.Jones, 1976. Interchange heterozygotes versus homozygotes. *Heredity* **39**:109–112.

Bailey R.L. and T.R.Dell, 1973. Quantifying diameter distributions with the Weibull function. *For. Sci.* **19**:97–104.

Baker F.S. 1925. Aspen in the Central Rocky Mountain region. *USDA Bull.* **1291**.

Baker F.S. 1949. A revised tolerance table. *J. For.* **56**:751–753.

Baker H.G. 1955. Self-compatibility and establishment after 'long-distance' dispersal. *Evolution* **9**:347–349.

Baker H.G. 1959. Reproductive methods as factors in speciation in flowering plants. *Cold Spring Harbor Symp. Quant. Biol.* **24**:177–190.

Baker H.G. 1965. Characteristics and modes of origin of weeds. In Baker H.G. and G.L.Stebbins (eds.), *The Genetics of Colonizing Species*, pp. 147–172. New York, Academic Press.

Baker H.G. 1966. The evolution, functioning and breakdown of heteromorphic incompatibility systems. *Evolution* **20**:349–368.

Baker H.G. 1967. Support for Baker's Law as a rule. *Evolution* **21**:853–856.

Baker H.G. 1972. Seed weight in relation to environmental conditions in California. *Ecol.* **53**:997–1010.

Baker H.G. 1974. The evolution of weeds. *Ann. Rev. Ecol. Syst.* **5**:1–24.

Baker H.G. and G.L.Stebbins (eds.), 1965. *The Genetics of Colonizing Species.* New York, Academic Press.

Bannister B.A. 1970. Ecological life cycle of *Euterpe globosa* Gaertn. In Odum H.T. and R.F.Pigeon (eds.), *A Tropical Rain Forest*, pp. B79–B89. Oak Ridge, Tenn., U.S. Atomic Energy Comm.

Bannister M.H. 1965. Variation in the breeding system of *Pinus radiata*. In Baker H.G. and G.L.Stebbins (eds.), *The Genetics of Colonizing Species*, pp. 353–372. New York, Academic Press.

Barker S.B., G. Cumming, and K.Horsfield, 1973. Quantitative morphometry of the branching structure of trees. *J. Theor. Biol.* **40**:33–43.

Barrett P.H. 1977. *The Collected Papers of Charles Darwin.* Chicago, University of Chicago Press.

Bartholomew B. 1970. Bare zone between California shrub and grassland communities: the rôle of animals. *Science* **170:**1210–1212.

Barton L.V. 1961. *Seed Preservation and Longevity.* London, Leonard Hill.

Baskin J.M. and C.C.Baskin, 1972. Germination characteristics of *Diamorpha cymosa* seeds and an ecological interpretation. *Oecologia* **10:**17–28.

Baskin J.M. and C.C.Baskin, 1973. Delayed germination in seeds of *Phacelia dubia* var. *dubia. Can. J. Bot.* **51:**2481–2486.

Baskin J.M. and C.C.Baskin, 1975. Ecophysiology of seed dormancy and germination in *Torilis japonica* in relation to its life cycle strategy. *Bull. Torr. Bot. Club* **102:** 67–72.

Bateman A.J. 1956. Cryptic self-incompatibility in the wallflower: *Chieranthus chieri* L. *Heredity* **10:**257–261.

Bawa K.S. 1974. Breeding systems of tree species of a lowland tropical community. *Evolution* **28:**85–92.

Bawa K.S. and P.A.Opler, 1975. Dioecism in tropical forest trees. *Evolution* **28:**85–92.

Bazzaz F.A. and J.L.Harper, 1977. Demographic analysis of the growth of *Linum usitatissimum. New Phytol.* **78:**193–208.

Beatley J.C. 1967. Survival of winter annuals in the northern Mohave desert. *Ecol.* **48:**745–750.

Beatley J.C. 1974. Phenological events and their environmental triggers in Mojave Desert ecosystems. *Ecol.* **55:**856–863.

Beattie A.J. 1976. Plant dispersion, pollination and gene flow in *Viola. Oecologia* **25:**291–300.

Beattie A.J. and N.Lyons, 1975. Seed dispersal in *Viola:* adaptations and strategies. *Am. J. Bot.* **62:**714–722.

Beddows A.R. 1931. Seed setting and flowering in various grasses. *Bull. Welsh Plant Breed. Stn. Series H,* **12:**5–99.

Beevers, L., D. Flescher, and R.H. Hageman. 1967. Studies of the pyridine nucleotide specificity of nitrate reductase in higher plants and its relationship to sulfhydryl level. *Biochem. Biophys. Acta* **89:**453–464.

Beg A., D.A.Emery and J.C.Wynne, 1925. Estimation and utilization of inter-cultivar competition in peanuts. *Crop Sci.* **15:**633–637.

Beisleigh W.J. and G.A.Yarranton, 1974. Ecological strategy and tactics of *Equisetum syl-* *vaticum* during a post-fire succession. *Can. J. Bot.* **52:**2299–2318.

Bell A.D. 1974. Rhizome organization in relation to vegetative spread in *Medeola Virginiana. J. Arnold Arb.* **55:**458–468.

Bell A.D. 1976. Computerized vegetative mobility in rhizomatous plants. In Lindenmayer A. and G.Rozenberg (eds.), *Automata, Languages, Development,* pp. 3–14. Amsterdam, North Holland.

Bell C.R. 1970. Seed distribution and germination experiment. In Odum H.T. and R.F.Pigeon (eds.), *A Tropical Rain Forest,* pp. D177–D182. Oak Ridge, Tenn., U.S. Atomic Energy Comm.

van den Bergh J.P. and C.T.deWit, 1960. Concurrentie tussen Timothee en Reukgras. *Meded. Inst. Biol. Scheik. Onderz. Landb. Gewass.* **121:**155–165.

Bernard J.M. and J.G.MacDonald, Jr. 1974. Primary production and life history of *Carex lacustris. Can. J. Bot.* **52:**117–123.

Bhatt G.M. 1973. Significance of path coefficient analysis in determining the nature of character association. *Euphytica* **22:**338–343.

Bhat J.L. 1973. Ecological significance of seed size to emergence and dormancy characteristics in *Indigofera glandulosa. Jap. J. Ecol.* **23:**95–99.

Bieleski R.L. 1959. Factors affecting growth and distribution of kauri (*Agathis australis* Salisb.). *Aust. J. Bot.* **7:**268–278.

Bingham J. 1969. The physiological determinants of grain yield in cereals. *Agric. Prog.* **44:**30–42.

Bingham J. 1971. Physiological objectives in breeding for grain yield in wheat. *Proc. 6th Eucarpia Congress* 15–29.

Black J.N. 1956. The influence of seed size and depth of sowing on pre-emergence and early vegetative growth of subterranean clover (*Trifolium subterraneum* L.). *Aust. J. agric. Res.* **7:**98–109.

Black J.N. 1958. Competition between plants of different initial seed sizes in swards of subterranean clover (*Trifolium subterraneum* L.) with particular reference to leaf area and the light microclimate. *Aust. J. agric. Res.* **9:**299–318.

Black J.N. 1961. Competition between two varieties of *Trifolium subterraneum* as related to the proportions of seed sown. *Aust. J. agric. Res.* **12:**810–820.

Black J.N. 1963. Defoliation as a factor in the growth of varieties of subterranean clover when grown in pure and mixed stands. *Aust. J. agric. Res.* **14:**206–225.

Black J.N. and G.N.Wilkinson, 1963. The role of time of emergence in determining the growth of individual plants in swards of subterranean clover. *Aust. J. agric. Res.* **14:**623–638.

Blackman G.E. and W.G.Templeman, 1938. The nature of competition between cereal crops and annual weeds. *J. agric. Sci., Camb.* **28:**247–271.

Blaisdell J.P. 1953. Ecological effects of planned burning of sagebrush-grass range on the upper Snake River plains. *U.S. Dept. Agr., Tech. Bull.* **1075.**

Bleasdale J.K.A. 1966. Plant growth and crop yield. *Ann. appl. Biol.* **57:**173–182.

Bliss C.I. and K.A.Reinker, 1964. A lognormal approach to diameter distributions in even-aged stands. *For. Sci.* **10:**350–360.

Blum B.M. 1961. Age–size relationships in all-aged northern hardwoods. *USDA Northeast. For. Expt. Stn., For. Res. Note* **125.**

Bocquet G. 1968. Cleistogamie et evolution chez les *Silene* L. sect. *Physolychnis* (Benth.) Bocquet (Caryophyllaceae) *Candollea* **23:**67–80.

Boisen A.T. and J.A.Newlin, 1910. The commercial hickories. *USDA For. Serv. Bull.* **80.**

Bond D.A. and M.Pope, 1947. Factors affecting the proportions of cross-bred and self-bred seed obtained from field bean (*Vicia faba* L.) crops. *J. agric. Sci.* **83:**343–351.

Bonner J.T. 1958. The relation of spore formation to recombination. *Amer. Nat.* **92:**193–200.

Boorman S.A. and P.R.Levitt, 1972. Group selection on the boundary of a stable population. *Theor. Pop. Biol.* **4:**85–128.

Bradbury I.K. and G.Hofstra, 1976. The partitioning of net energy resources in two populations of *Solidago canadensis* during a single developmental cycle in southern Ontario. *Can. J. Bot.* **54:**2449–2456.

Bradley R. 1721. *A Philosophical Account of the Works of Nature.* London, Mears.

Bradshaw A.D. 1972. Some of the evolutionary consequences of being a plant. *Evolutionary Biology* **5:**25–47.

Braid K.W. 1948. Bracken control—artificial and natural. *J. Brit. Grassl. Soc.* **3:**181–189.

Branson F.A. 1955. Relationships of grass morphology to grazing resistance. *Proc. Montana Acad. Sci.* **15:**41–43.

Braun A. 1853. Das Individuum der Pflanze in seinem Verhältniss zur Species-Generationsfolge, Generationswechsel und Generationstheilung der Pflanze. *Abh. Kgl. Preuss. Akad. Wissensch. Berlin,* 19–122. (English trans. in *Amer. J. Arts. Sci.* 2nd Ser.,

19:297–318 (1855); **20:**181–201 (1855); **21:**58–79 (1856); also in *Ann. Nat. Hist.* **16:**233–256, 333–354 (1855); **18:**363–386 (1857).

Brenchley W.E. 1918. Buried weed seeds. *J. agric. Sci.* **9:**1–31.

Brenchley W.E. and K.Warington, 1930. The weed seed population of arable soil. I. Numerical estimation of viable seeds and observations on their natural dormancy. *J. Ecol.* **18:**235–272.

Brenchley W.E. and K.Warington, 1933. The weed seed population of arable soil. II. Influence of crop, soil, and methods of cultivation upon the relative abundance of viable seeds. *J. Ecol.* **21:**103–127.

Briscoe C.B. and F.H.Wadsworth, 1970. Stand structure and yield in the tabonuco forest of Puerto Rico. In Odum H.T. and R.F.Pigeon (eds.), *A Tropical Rain Forest,* pp. B79–B89. Oak Ridge, Tenn., U.S. Atomic Energy Comm.

Brix H. 1967. An analysis of dry matter production of Douglas fir seedlings in relation to temperature and light intensity. *Can. J. Bot.* **45:**2063–2072.

Brncic, D. 1954. Heterosis and the integration of the genotype in geographical populations of *Drosophila pseudoobscura. Genetics* **39:**77–88.

Brougham R.W. 1959. The effects of frequency and intensity of grazing on the productivity of a pasture of short-rotation ryegrass and red and white clover. *N.Z. J. agric. Res.* **2:**1232–1248.

Bucio Alanis L., J.M.Perkins and J.L.Jinks, 1969. Environmental and genotype-environmental components of variability. V. Segregating generations. *Heredity* **24:**115–127.

Bunting A.H. 1960. Some reflections on the ecology of weeds. In Harper J.L. (ed.), *The Biology of Weeds,* pp. 11–26. Oxford, Blackwell Scientific Publications.

Burgess P.F. 1970. An approach towards a silvicultural system for the hill forests of the Malay Peninsula. *Malay. Forester* **24:**66–80.

Burgess P.F. 1972. Studies on the regeneration of the hill forests of the Malay Peninsula. The phenology of Dipterocarps. *Malay. Forester* **35:**103–123.

Burrows F.M. 1973. Calculation of the primary trajectories of plumed seeds in steady winds with variable convection. *New Phytol.* **72:**647–664.

Burrows F.M. 1975. Wind-borne seed and fruit movement. *New Phytol.* **75:**405–418.

Büsgen E. and Münch E. 1929. *The Structure and Life of Forest Trees.* New York, Wiley.

Bythe D.E. and C.R.Weber, 1968. Effects of genetic heterogeneity within two soybean populations. I. Variability within environments and stability across environments. *Crop Sci.* **8**:44–47.

Caldwell P.A. 1957. The spatial development of *Spartina* colonies growing without competition. *Ann. Bot. (Lond.)* **21**:203–216.

Callaghan T.V. 1976. Strategies of growth and population dynamics of tundra plants. 3. Growth and population dynamics of *Carex bigelowii* in an alpine environment. *Oikos* **27**:402–413.

Callaghan T.V. and N.J.Collins, 1976. Strategies of growth and population dynamics of tundra plants. I. Introduction. *Oikos* **27**:383–388.

Carlquist S. 1966. The biota of long-distance dispersal. IV. *Evolution* **20**:433–455.

Carlquist S. 1967. The biota of long distance dispersal. V. Plant dispersal to Pacific islands. *Bull. Torrey Bot. Club* **94**:129–141.

Carlquist S. 1974. *Island Biology.* New York, Columbia University Press.

Carpenter F.L. 1976. Plant-pollinator interactions in Hawaii: pollination energetics of *Metrosideros collina (Myrtaceae)*. *Ecology* **57**:1125–1144.

Catlin C.N. 1925. Composition of Arizona forages with comparative data. *Ariz. Univ. Agr. Expt. Sta. Bull.* **113**:155–173.

Cavers P.B. and J.L.Harper, 1967. Studies in the dynamics of plant populations. I. The fate of seed and transplants introduced into various habitats. *J. Ecol.* **55**:59–71.

Champness S.S. and K.Morris. 1948. The population of buried viable weed seeds in relation to contrasting pasture and soil types. *J. Ecol.* **36**:149–173.

Chancellor R.J. 1966. Weed seeds in the soil. *Rep. A.R.C. Weed Res. Org. 1960–64,* 15–19.

Chancellor R.J. 1974. The development of dominance among shoots arising from fragments of *Agropyron repens* rhizomes. *Weed Res.* **14**:29–38.

Chancellor R.J. and N.C.B.Peters, 1974. The time of onset of competition between wild oats and spring cereals. *Weed Res.* **14**:197–202.

Charles A.H. 1961. Differential survival of cultivars of *Lolium, Dactylis* and *Phleum. J. Br. Grassl. Soc.* **16**:69–75.

Charles A.H. 1970. Ryegrass populations from intensively managed leys. I. Seedling and spaced plant characters. *J. agric. Sci., Camb.* **75**:103–107.

Chippindale H.G. and W.E.J.Milton, 1934. On the viable seeds present in the soil beneath pastures. *J. Ecol.* **22**:508–531.

Christensen N.L. and C.H.Müller, 1975. Relative importance of factors controlling germination and seedling survival in *Adenostema* chaparral. *Amer. Nat.* **93**:71–78.

Christiansen F.B. and O.Frydenberg, 1973. Selection component analysis of natural polymorphisms using population samples including mother–offspring combinations. *Theor. Pop. Biol.* **4**:425–445.

Christiansen F.B. and O. Frydenberg, 1976. Selection component analysis of natural polymorphisms using mother–offspring samples of successive cohorts. In Karlin S. and E.Nevo (eds.), *Population Genetics and Ecology,* pp. 277–303. New York, Academic Press.

Clausen J. 1926. Genetical and cytological investigations on *Viola tricolor* L. and *V. arvensis* Murr. *Hereditas* **8**:1–156.

Clausen J. and W.M.Hiesey, 1958. Experimental studies on the nature of species. IV. Genetic structure of ecological races. *Carnegie Inst. Washington. Pub.* **615**.

Clausen J., D.D.Keck and W.M.Hiesey, 1940. Experimental studies on the nature of species. I. Effect of varied environments on western North American plants. *Carnegie Inst. Wash. Publ.* **520**.

Clausen J., D.D.Keck and W.M.Hiesey, 1947. Heredity of geographically and ecologically isolated races. *Amer. Nat.* **81**:114–133.

Clausen R.E. and D.R.Cameron, 1950. Inheritance in *Nicotiana tabacum.* XXII. Duplicate factors for chlorophyll production. *Genetics* **35**:4–10.

Clegg M.T. 1979. Genetic demography of plant populations (in press).

Clegg M.T. and R.W.Allard, 1972. Patterns of genetic differentiation in the slender wild oat species *Avena barbata. Proc. Nat. Acad. Sci. USA* **69**:1820–1824.

Clegg M.T., R.W. Allard, and A.L. Kahler. 1972. Is the gene the unit of selection? Evidence from two experimental populations. *Proc. Natl. Acad. Sci. U.S.* **69**:2474–2478.

Clegg M.T., A.L.Kahler and R.W.Allard, 1978. Estimation of life cycle components of selection in an experimental plant population. *Genetics* **89**:765–792.

Clements F.E. 1920. Adaptation and mutation as a result of fire. *Carnegie Inst. Wash., Yearbook* **19**:348–349.

Cohen D. 1966. Optimizing reproduction in a randomly varying environment. *J. Theor. Biol.* **12:**119–129.

Cohen D. 1968. A general model of optimal reproduction in a randomly varying environment. *J. Ecol.* **56:**219–228.

Connell J.H. 1971. On the role of natural enemies preventing competitive exclusion in some marine animals and in forest trees. In den Boer P.J. and G.R.Gradwell (eds.), *The Dynamics of Populations*, pp. 298–310. Oosterbeek, Netherlands.

Cook R.E. 1975. The photoinductive control of seed weight in *Chenopodium rubrum* L. *Am. J. Bot.* **62:**427–431.

Cook R.E. 1976. Photoperiod and the determination of potential seed number in *Chenopodium rubrum* L. *Ann. Bot.* **40:**1085–1099.

Cook R.E. 1979. Patterns of juvenile mortality and recruitment in plants. In Solbrig O.T., S.Jain, G.Johnson and P.H.Raven (eds.), *Topics in Plant Population Biology*, New York, Columbia University Press.

Coombe D.E. 1960. An analysis of the growth of *Trema guineensis*. *J. Ecol.* **48:**219–231.

Coombe D.E. and W.Hadfield, 1962. An analysis of the growth of *Musanga cecropioides*. *J. Ecol.* **50:**221–234.

Corner E.J.H. 1954. The evolution of tropical forest. In Huxley J.S., A.C.Hardy and E.B.Ford (eds.), *Evolution as a Process*, pp. 34–46. London, Allen and Unwin.

Corner E.J.H. 1964. *The Life of Plants*, Cleveland, Ohio, World Publishing Co.

Cottam W.P. 1954. Prevernal leafing of aspen in Utah Mountains. *J. Arnold Arb.* **35:**239–250.

Coyne J.A. 1976. Lack of genic similarity between two sibling species of *Drosophila* as revealed by varied techniques. *Genetics* **84:**593–607.

Crane M.B. and W.J.C.Lawrence. 1934. *The Genetics of Garden Plants*. London, Macmillan.

Cress C.E. 1966. Heterosis of the hybrid related to gene frequency differences between two populations. *Genetics* **53:**269–274.

Croat T. 1969. Seasonal flowering behaviour in Central Panama. *Ann. Mo. Bot. Gard.* **56:**295–307.

Crocker W. 1938. Life-span of seeds. *Bot. Rev.* **4:**235–274.

Crow T.R. 1977. A rain forest chronicle: a thirty year record of change in structure and composition at El Verde, Puerto Rico (unpublished MS).

Crumpacker D.W. 1967. Genetic loads in maize (*Zea mays* L.) and other cross-fertilized plants and animals. *Evol. Biol.* **1:**306–424.

Curran P.L. 1963. Balanced polymorphism in *Dactylis glomerata* sub-species *woronowii*. *Nature* (London). **197:**105–106.

Cussans G.W. 1974. The biological contribution to weed control. In Price-Jones D. and M.E.Soloman (eds.), *Biology in Pest and Disease Control*. Oxford, Blackwell Scientific Publications.

Cussans G.W. 1976. Population studies. In Price-Jones D. (ed.), *Wild Oats in World Agriculture*, pp. 119–125. London, Agricultural Res. Council.

Daniel G.H. 1955. Dredge corn trials 1946–51. *J. Nat. Inst. agric. Bot.* **7:**309–317.

Dansereau P. 1961. The origin and growth of plant communities. In Zarrow M.X. (ed.), *Growth in Living Systems*, pp. 567–603. New York, Basic Books.

Dansereau P. 1971. The variety of coenotypes in vascular plants and the spectrum of their distribution in several communities. 1. Definition and test. *Natur. Can.* **98:**359–384.

Darlington C.D. 1939. *Evolution of Genetic systems*. Edinburgh, Oliver and Boyd.

Darwin C. 1839. *Journal of Researches into the Natural History and Geology of the countries visited during the voyage of H.M.S. Beagle round the world under the command of Capt. Fitzroy, R.N.* London, Colburn.

Darwin C. 1859. *The Origin of Species*. London, Murray.

Darwin C. 1876. *The Effects of Cross and Self-fertilisation in the Vegetable Kingdom*. London, Murray.

Darwin C. 1877. *The Different Forms of Flowers on Plants of the same Species*. London, Murray.

Darwin E. 1800. *Phytologia*. London.

Daubenmire R. 1968. *Plant Communities. A Textbook of Plant Synecology*. New York, Harper & Row.

Davies M.S. and R.W.Snaydon, 1973. Physiological differences among populations of *Anthoxanthum odoratum*. I. Response to calcium. *J. Appl. Ecol.* **10:**33–45.

Davies M.S. and R.W.Snaydon, 1976. Rapid population differentiation in a mosaic environment. III. Measures of selection pressures. *Heredity* **36:**59–66.

Davies W.E. and N.R.Young, 1966. Self-fertility in *Trifolium fragiferum*. *Heredity* **21:**615–624.

Dawkins, H.C. 1956. Rapid growth of aberrant girth increment of rain forest trees. *Emp. for. Rev.* **35**:449–454.

Dawkins H.C. 1966. The time dimension of tropical forest trees. *J. Ecol.* **53**:837P–838P.

DeBenedictis P.A. 1978. Are populations characterized by their genes or by their genotypes. *Amer. Nat.* **112**:155–175.

Deevey E.S. 1947. Life tables for natural populations of animals. *Quart. Rev. Biol.* **22**:283–314.

deLiocourt F. 1898. De l'amenagement des sapinières. *Bull. Soc. For. France-Conté et Belfort* **4**:396–409, 645–647.

Dennis F.G., J.Lipecki and C.Kiang, 1970. Effects of photoperiod and other factors upon flowering and runner development of three strawberry cultivars. *J. Amer. Soc. Hort. Sci.* **95**:750–754.

Dew D.A. 1972. An index of competition for estimating crop loss due to weeds. *Can. J. Pl. Sci.* **52**:921–927.

Dewsberry R. 1977. Plant self-thinning dynamics. *Planta* **136**:249–252.

Dobzhansky Th. 1959. Evolution of genes and genes in evolution. *Cold Spring Harbor symp. Quant. Biol.* **24**:15–30.

Dobzhansky Th. 1970. *Genetics of the Evolutionary Process*. New York, Columbia University Press.

Donald C.M. 1963. Competition among crop and pasture plants. *Adv. Agron.* **15**:1–118.

Donald C.M. 1968. The breeding of crop ideotypes. *Euphytica* **17**:385–403.

Doney D.L., R.L.Plaisted and L.C.Peterson, 1966. Genotypic competition in progeny performance evaluation of potatoes. *Crop Sci.* **6**:433–435.

Douglas D.A. 1977. The Reproductive Strategy of *Mimulus primuloides* Benth. (Scrophulariaceae) Ph.D. Thesis. University of California (Berkeley).

Drew T.J. and F.Lewelling, 1977. Some recent Japanese theories of yield-density relationships and their application to Monterey pine plantations. *For. Sci.* **23**:517–534.

Duke S.O., G.H.Egley and B.J.Reger, 1977. Model for variable light sensitivity in imbibed dark-dormant seeds. *Plant Physiol.* **59**:244–249.

Eagles C.F. 1972. Competition for light and nutrients between natural populations of *Dactylis glomerata. J. Appl. Ecol.,* **9**:141–151.

East E.M. 1916. Studies on size inheritance in *Nicotiana. Genetics* **1**:164–176.

Eberhart S.A., L.H.Penny and G.F.Sprague, 1964. Intra-plot competition among maize single-crosses. *Crop Sci.* **4**:467–471.

Ehrlich P.R. and P.H. Raven, 1969. Differentiation of populations. *Science.* **165**:1228–1232.

Ek A.R. and D.H.Dawson, 1976. Actual and projected growth and yields of *Populus 'Tristis #1'* under intensive culture. *Can. J. For. Res.* **6**:132–144.

Ellis W.M. 1974. The breeding system and variation in *Poa annua* L. *Evolution* **27**:656–662.

Emerson R.A. 1921. The genetic relations of plant colours in maize. *Cornell Univ. Agr. Exp. Sta. Mem.* **39**.

Emerson R.A., G.W. Beadle and A.C.Fraser. 1935. A summary of linkage studies in maize. *Cornell Univ. Agr. Exp. Sta. Mem.* **80**.

England F. 1967. Non-sward densities for the assessment of yield in Italian ryegrass. I. Comparison between swards and non-sward densities. *J. agric. Sci., Camb.* **68**:235–241.

England F.C. 1968. Competition in mixtures of herbage grasses. *J. Appl. Ecol.* **5**:227–242.

Epling C., H.Lewis and F.M.Ball, 1960. The breeding group and seed storage: a study in population dynamics. *Evol.* **14**:238–255.

Ernst A. 1953. 'Basic numbers' und polyploidy und ihre Bedeutung für des Heterostylieproblem. *Arch. Klaus-Stift. Verebf.* **28**:1–159.

Evans L.T. 1976. Physiological adaptation to performance as crop plants. *Phil. Trans. R. Soc. Lond. b.* **275**:71–83.

Evans L.T. and I.F.Wardlaw, 1976. Aspects of the comparative physiology of grain yield in cereals. *Adv. Agron.* **28**:301–359.

Faegri K. and L.van der Pijl, 1971. *The Principles of Pollination Ecology*. Toronto, Pergamon Press.

Fedorov A.A. 1966. The structure of the tropical rain forest and speciation in the humid tropics. *J. Ecol.* **54**:1–11.

Fehr W.R. 1973. Evaluation of intergenotypic competition with a paired-row technique. *Crop Sci.* **13**:572–575.

Felton W.L. 1976. The influence of row spacing and plant population on the effects of weed competition in soybean. *Aust. J. Exp. Agric. Husb.* **16**:926–931.

Fergus E.N. 1922. Self-fertility in red clover. *Kentucky Aric. Exp. Sta. Circ. no. 29.*

Finnerty V. and G.B.Johnson, 1979. Post translational modification as a potential explanation of high levels of enzyme polymorphism: xanthine dehydrogenase and aldehyde oxidase in *Drosophila melanogaster*. *Genetics* (In press).

Fisher R.A. 1941. Average excess and average effect of a gene substitution. *Annals Eugenics* **11**:53–63.

Flower-Ellis J.G.K. 1971. Age structure and dynamics in stands of bilberry (*Vaccinium myrtillus* L.). *Rapp. Uppsats. Avdel. Skogsekol.* **9.**

Fournier L.A. and S.Salas, 1967. Tabla de vida para el primer año de la población de *Dipterodendron costarricense* Radlk. *Turrialba* **17**:348–350.

Fox J.E.D. 1972. *The natural vegetation of Sabah and natural regeneration of the dipterocarp forests*. Ph.D. thesis, University of Wales.

Frankie G.W. 1975. Tropical forest phenology and pollinator plant coevolution. In Gilbert L.E. and P.H.Raven (eds.), *Coevolution of Animals and Plants*, pp. 192–209. Austin, University of Texas Press.

Frankie G.W., H.G.Baker and P.A.Opler, 1974. Comparative phenological studies of trees in tropical lowland wet and dry forest sites of Costa Rica. *J. Ecol.* **62**:881–919.

Frankie G.W., P.A.Opler and K.S.Bawa, 1976. Foraging behaviour of solitary bees: implications for outcrossing of a neotropical forest tree species. *J. Ecol.* **64**:1049–1057.

Free J.B. 1970. *Insect Pollination of Crops*. New York, Academic Press.

Free J.B. and I.H.Williams, 1976. Pollination as a factor limiting the yield of field beans (*Vicia faba* L.). *J. agric. Sci.* **87**: 395–399.

Frey K.J. and U.Maldonado, 1967. Relative productivity of homogeneous and heterogeneous oat cultivars in optimum and suboptimum environments. *Crop Sci.* **7**:532–553.

Frothingham E.H. 1912. Second-growth hardwoods in Connecticut. *USDA For. Serv. Bull.* **96.**

Gadgil M. and O.T.Solbrig, 1972. The concept of *r*- and *K*-selection: Evidence from wild flowers and some theoretical considerations. *Amer. Nat.* **106**:14–31.

Gaertner J. 1788–1791. *De Fructibus et Seminibus Plantarum*. Stuttgart.

Gaines M.S., K.J.Vogt, L.Hamrick and J.Caldwell, 1974. Reproductive strategies and growth patterns in sunflowers (*Helianthus*). *Amer. Nat.* **108**:889–894.

Ganders F.R. 1975a. Heterostyly, homostyly, and fecundity in *Amsinckia spectabilis* (Boraginaceae). *Madroño* **23**:56–62.

Ganders F.R. 1975b. Mating patterns in self-compatible distylous populations of *Amsinckia* (Boraginaceae). *Can. J. Bot.* **53**:773–779.

Garnock-Jones P.J. 1976. Breeding systems and pollination in New Zealand *Parahebe* (Scrophulariaceae). *N.Z. J. Bot.* **14**: 291–298.

Garwood E.A. 1969. Seasonal tiller populations of grass and grass/clover swards with and without irrigation. *J. Brit. Grassl. Soc.* **24**:333–344.

Gates F.C. and G.E.Nichols, 1930. Relation between age and diameter in trees of the primeval northern hardwood forest. *J. For.* **28**:395–398.

Georghiou G.P. 1972. The evolution of resistance to pesticides. *Ann. Rev. Ecol. Syst.* **3**:133–168.

Gibbs C.B. 1963. Tree diameter a poor predictor of age in West Virginia hardwoods. *USDA For. Serv. Res. Note* **NE-11.**

Gill A.M. 1969. The ecology of an elfin forest in Puerto Rico. 6. Aërial roots. *J. Arnold Arb.* **50**:197–209.

Gilmartin, A.J. 1968. Baker's Law and dioecism in the Hawaiian Flora: an apparent contradiction. *Pacific Science* **22**:285–292.

Givnish T. 1979. On the Adaptive Significance of Leaf Form. In Solbrig O.T., S.Jain, G.Johnson and P.H.Raven (eds.), *Topics in Plant Population Biology*. New York, Columbia University Press.

Glasgow J.L., J.W.Dick and D.R.Hodgson, 1976. Competition by, and chemical control of, natural weed populations in spring-grown field beans. *Ann. Appl. Biol.* **84**:259–269.

Glendening G.E. 1952. Some quantitative data on the increase of mesquite and cactus on a desert grassland range in southern Arizona. *Ecology* **33**:319–328.

Glenn F.B. and T.B.Daynard, 1974. Effects of genotype, planting pattern and plant density on plant-to-plant variability and grain yield of corn. *Can. J. Pl. Sci.* **54**:323–330.

Godwin H. 1968. Evidence for longevity of seeds. *Nature* **220**:708–709.

Godwin H. 1975. *The History of the British Flora*. Cambridge, Cambridge University Press.

Goel N.S. and N.Richter-Dyn, 1974. *Stochastic Models in Biology*. New York, Academic Press.

Goethe J.W. 1790. *Versuch die Metamorphose der Pflanzen zu erklären*. (English translation by Arber A. 1946, 'Goethe's Botany'. *Chronica Botanica* **10**(2)).

Goff F.G. and D.West, 1975. Canopy-understory interaction effects on forest population structure. *For. Sci.* **21**:98–108.

Goodall D.W. 1949. A quantitative study of the early development of the seedling of Cacao *(Theobroma cacao)*. *Ann. Bot. N.S.* **13**:1–21.

Goodall D.W. 1955. Growth of cacao seedlings as affected by illumination. *Rep. XIV Int. Hort. Cong.* Wageningen, Netherlands.

Gorham E. and M.G.Somers, 1973. Seasonal changes in the standing crop of two montane sedges. *Can. J. Bot.* **51**:1097–1108.

Gottlieb L.D. 1974. Genetic stability in a peripheral isolate of *Stephanomeria exigua sp. coronaria* that fluctuates in population size. *Genetics* **76**:551–556.

Grant V. 1958. The regulation of recombination in plants. *Cold Spring Harbor Symp. Quant. Biol.* **23**:337–363.

Grant V. 1963. *The Origin of Adaptations.* New York, Columbia University Press.

Grant V. 1964. *The Architecture of the Germplasm.* New York, John Wiley.

Grant V. 1975. *Genetics of Flowering Plants.* New York, Columbia University Press.

Grant V. and K.A.Grant, 1965. *Flower Pollination in the Phlox Family.* New York, Columbia University Press.

Green J.O. and J.C.Eyles, 1960. A study in methods of grass variety testing. *J. Brit. Grassl. Soc.* **15**:124–132.

Grime J.P. 1966. Shade avoidance and tolerance in flowering plants. In Brainbridge R., E.C.Evans and D.Rackhan (eds.) *Light As An Ecological Factor,* pp. 281–301. Oxford, Blackwell Scientific Publications.

Grime J.P. 1974. Vegetation classification by reference to strategies. *Nature* **234**:96–97.

Grime J.P. 1977. Evidence for the existence of three primary strategies in plants and its relevance to ecological and evolutionary theory. *Amer. Nat.* **111**:1169–1194.

Grime J.P. and R.Hunt, 1975. Relative growth rate: its range and adaptive significance in a local flora. *J. Ecol.* **63**:393–422.

Grimes D.W. and J.T.Musick, 1960. Effect of plant spacing fertility and irrigation management on grain sorghum production. *Agron. J.* **52**:647–650.

Guevara F. 1977. *Estudios sobre la dinámica de poblaciones de semillas de* Cordia elaeagnoides, *en Chamela, Jal.* Tesis, Fac. Ciencias, UNAM, México.

Guevara S. and A.Gómez-Pompa, 1972. Seeds from surface soils in a tropical region of Veracruz, México. *J. Arnold Arb.* **53**:312–335.

Gustafsson A. 1946. Apomixis in the higher plants. I. The mechanism of apomixis. *Lunds Univ. Arsskr. N. F. Aud.* 2, **42**:1–66.

Gustafsson A. 1947a. Apomixis in the higher plants. II. The causal aspect of apomixis. *Lunds Univ. Arsskr. N. F. Aud.* 2, **43**:71–178.

Gustafsson A. 1947b. Apomixis in the higher plants. III. Biotype and species formation. *Lunds Univ. Arsskr. N. F. Aud.* 2, **44**:183–370.

Hageman R.H., E.R. Leng and J.W.Dudley, 1967. A biochemical approach to corn breeding. *Adv. in Agron.* **19**:45–86.

Hagerup O. 1932. On pollination in the extremely hot air at Timbuctu. *Dansk. Bot. Ark.* **8**:1–20.

Hagerup O. 1951. Pollination in the Faroes—in spite of rain and poverty in insects. *D. Kgl. danske Vidensk. Selsk. Biol. Medd.* **18**, No. 15:1–47.

Haig I.T. 1932. Second-growth yield, stand and volume tables for the Western White Pine type. *USDA Tech. Bull.* **323**.

Hallé F. and R.A.A.Oldeman, 1970. *Essai sur l'Architecture et la Dynamique de Croissance des Arbres Tropicaux.* Paris, Masson.

Hallé F., R.A.A.Oldeman and P.B.Tomlinson. 1978. *Tropical Trees and Forests: An Architectural Analysis.* Berlin, Springer.

Halloran G.M. and W.J. Collins, 1974. Physiological predetermination of the order of hardseededness breakdown in subterranean clover *(Trifolium subterraneum* L.). *Ann. Bot.* **38**:1039–1043.

Hamlin J. and C.M. Donald, 1974. The relationship between plant form, competitive ability and grain yield in barley. *Euphytica* **23**:535–542.

Hamrick J.L. 1976. Variation and selection in western montane species. II. Variation within and between populations of white fir on an elevational transect. *Theoret. Appl. Genet.* **47**:27–34.

Hamrick J.L. 1979. Genetic variation and longevity. In Solbrig O.T., S.Jain, G.Johnson and P.H.Raven (eds), *Topics in Plant Population Biology.* New York, Columbia University Press.

Hamrick J.L. and R.W.Allard, 1972. Microgeographical variation in allozyme frequencies in *Avena barbata. Proc. Nat. Acad. Sci. U.S.* **69**:2100–2104.

Hamrick J.L. and R.W.Allard, 1975. Correlations between quantitative characters and enzyme genotypes in *Avena barbata. Evolution.* **29**:438–442.

Hancock J.F. and R.E.Wilson, 1976. Biotype selection in *Erigeron annuus* during old field succession. *Bull. Torrey Bot. Club.* **103**:122–125.

Hanes T.L. 1971. Succession after fire in the chaparral of southern California. *Ecol. Monogr.* **41**:27–52.

Harberd D.J. 1961. Observations on population structure and longevity of *Festuca rubra* L. *New Phytol.* **60**:184–206.

Harberd D.J. 1962. Some observations on natural clones in *Festuca ovina. New Phytol.* **61**:81–100.

Harberd D.J. 1967. Observations on natural clones of *Holcus mollis. New Phytol.* **66**:401–408.

Harding J., R.W.Allard and D.G.Smeltzer, 1966. Population studies in predominately self-pollinating species. IX. Frequency-dependent selection in *Phaseolus lunatus. Proc. Nat. Acad. Sci.* **56**:99–104.

Harlan H.V. and M.L.Martini, 1938. The effects of natural selection in a mixture of barley varieties. *J. Agric. Res.* **57**:189–99.

Harland S.C. 1936. The genetical conception of the species. *Biol Rev.* **11**:83–112.

Harms W.R. and O.G. Langdon, 1976. Development of loblolly pine in dense stands. *For. Sci.* **22**:331–337.

Harper J.L. 1957. Biological flora of the British Isles: *Ranunculus repens, R. acris, R. bulbosa. J. Ecol.* **45**:289–342.

Harper J.L. 1959. The ecological significance of dormancy and its importance in weed control. *Proc. 4th Int. Congr. Crop Prot.* 415–520.

Harper J.L. 1961. Approaches to the study of competition. *Symp. Soc. Exp. Biol.* **15**:1–39.

Harper J.L. 1965. The nature and consequences of interference amongst plants. *Proc. XI Internat. Cong. Genet.* **2**:465–482.

Harper J.L. 1967. A Darwinian approach to plant ecology. *J. Ecol.* **55**:247–270.

Harper J.L. 1969. The role of predation in vegetational diversity. *Brookhaven Nat. Lab. Symp. Biol.* **22**:48–62.

Harper J.L. 1977. *Population Biology of Plants.* London, Academic Press.

Harper J.L. and R.A.Benton, 1966. The behavior of seeds in soil. II. The germination of seeds on the surface of a water-supplying substrate. *J. Ecol.* **54**:151–166.

Harper J.L. and D.Gajic, 1961. Experimental studies of the mortality and plasticity of a weed. *Weed Res.* **1**:91–104.

Harper J.L., P.H.Lovell and K.G.Moore, 1970. The shapes and sizes of seeds. *Ann. Rev. Ecol. Syst.* **1**:327–356.

Harper J.L. and J.Ogden, 1970. The reproductive strategy of higher plants. I. The concept of strategy with special reference to *Senecio vulgaris* L. *J. Ecol.* **58**:681–698.

Harper J.L. and J.White, 1971. The dynamics of plant populations. *Proc. Advanced Study Inst. Dyn. Numbers Pop., Oosterbeek 1970,* pp. 41–63.

Harper J.L. and J.White, 1974. The demography of plants. *Ann. Rev. Ecol. Syst.* **5**:419–463.

Harper J.L., W.T.Williams and G.R.Sagar, 1965. The behavior of seeds in soil. I. The heterogeneity of soil surfaces and its role in determining the establishment of plants from seed. *J. Ecol.* **53**:273–286.

Harrington J.F. 1972. Seed storage and longevity. In Kozlowski T.T. (ed.), *Seed Biology,* pp. 145–245. New York, Academic Press.

Harris H. 1966. Enzyme polymorphism in man. *Proc. Roy. Soc.* (London) B, **164**:298–310.

Harris W. 1970. Competition effects on yield and plant and tiller density in mixtures of ryegrass cultivars. *Proc. N.Z. Ecol. Soc.* **17**:10–17.

Harris W. 1973. Competition among pasture plants. IV. Cutting frequency, nitrogen and phosphorus interactions, and competition between two ryegrass cultivars. *N.Z. J. Agric. Res.* **16**:399–413.

Harris W. and R.W.Brougham, 1968. Some factors affecting change in botanical composition in a ryegrass-white clover pasture under continuous grazing. *N.Z. J. Agric. Res.* **11**:15–38.

Harris W. and V.J.Thomas, 1970. Competition among pasture plants. I. Effects of frequency of cutting on competition between two ryegrass cultivars. *N.Z. J. Agric. Res.* **13**:833–861.

Hartnett D.C. and W.G.Abrahamson, 1979. The effects of stem gall insects on life history patterns in *Solidago canadensis* L. (Compositae). *Ecology.* (In press).

Hartshorn G.S. 1972. *Ecological life history and population dynamics of* Pentaclethra macroloba, *a tropical wet forest dominant, and* Stryphnodendron excelsum, *an occasional associate.* Ph.D. Thesis, University of Washington (Seattle), 118 pp.

Hartshorn G.S. 1975. A matrix model of tree population dynamics. In Golley F.B. and E.Medina (eds.), *Tropical Ecological Systems,* pp. 41–52. New York, Springer-Verlag (Ecological Studies 11).

Hartshorn G.S. 1977. Tree falls and tropical forest dynamics. In Tomlinson P.B. and M.H.Zimmermann (eds.), *Tropical Trees as Living Systems*. Cambridge, Cambridge University Press.

Haslam S.M. 1970. The development of the annual population in *Phragmites communis* Trin. *Ann. Bot.* **34**:571–591.

Hatcher R.J. 1963. A study of black spruce forests in northern Quebec. *Canad. Dept. For. Publ.* **1018**.

Hawthorn W.R. and P.B.Cavers, 1976. Population dynamics of the perennial herbs *Plantago major* and *P. rugelii*. *J. Ecol.* **64**:511–527.

Hayashi I. and M.Numata, 1971. Ecological studies on the buried seed population in the soil related to plant succession. IV. *Jap. J. Ecol.* **20**:243–252.

Hayward M.D. and E.L.Breese, 1966. The genetic organization of natural populations of *Lolium perenne*. I. Seed and seedling characters. *Heredity* **21**:287–304.

Hedrick P.W., M.E.Ginevan and E.P.Ewing, 1976. Genetic polymorphism in heterogeneous environments. *Ann. Rev. Ecol. Syst.* **7**:1–32.

Heinrich, B. 1975. Bee flowers: a hypothesis on flower variety and blooming times. *Evolution* **29**:325–334.

Heithaus E.R., P.A.Opler and H.G.Baker, 1974. Bat activity and pollination of *Bauhinia pauletia*: plant-pollinator coevolution. *Ecology* **55**:412–419.

Henry J.D. and J.M.A.Swan, 1974. Reconstructing forest history from live and dead plant material—an approach to the study of forest succession in southwest New Hampshire. *Ecol.* **55**:772–783.

Henslow G. 1879. On the self-fertilization of plants. *Linn. Soc. Trans. Ser. II. Bot.* **1**:317–398.

Hett J.M. and O.L.Loucks, 1976. Age structure models of balsam fir and eastern hemlock. *J. Ecol.* **64**:1029–1044.

Heslop-Harrison, J. 1964. Forty years of genecology. *Adv. Ecol. Res.* **2**:159–247.

Heydecker W. 1973. *Seed Ecology*. London, Butterworth.

Hickman J.C. 1979. The Basic Biology of Plant Numbers. In Solbrig O.T., S.Jain, G.Johnson and P.H.Raven (eds.), *Topics in Plant Population Biology*. New York, Columbia University Press.

Hicks D.R. and R.E.Stucker, 1972. Plant density effect on grain yield of corn hybrids diverse in leaf orientation. *Agron. J.* **64**:484–487.

Hinson K. and W.D.Hanson, 1962. Competition studies in soybeans. *Crop Sci.* **2**:117–123.

Hiroi T. and M. Monsi, 1966. Dry matter economy of *Helianthus annuus* communities grown at varying densities and light intensities. *J. Fac. Sci. Univ. Tokyo* **9**:241–285.

Hodgkinson K.C. 1970. The effects of frequency and extent of defoliation, summer irrigation and fertilizer on the production and survival of the grass *Danthonia caespitosa*. *Aust. J. Agric. Res.* **27**:755–767.

Holland P.G. 1969. The maintenance of structure and shape in three mallee eucalypts. *New Phytol.* **68**:411–421.

Holler L.C. and W.G.Abrahamson, 1977. Seed and vegetative reproduction in relation to density in *Fragaria virginiana* (Rosaceae). *Amer. J. Bot.* **64**:1003–1007.

Holliday R.J. 1960. Plant population and crop yield. *Field Crop Abstr.* **13**:159–167.

Holm T. 1925. Hibernation and rejuvenation, exemplified by North American herbs. *Amer. Midl. Nat.* **9**:439–512.

Holttum R.E. 1938. The ecology of tropical pteridophytes. In F. Verdoorn (ed.), *Manual of Pteridology*, pp. 420–450. The Hague, Nijhoff.

Horn H.S. 1971. *The Adaptive Geometry of Trees*. Princeton, Princeton University Press.

Horton K.W. and E.J.Hopkins, 1965. Influence of fire on aspen suckering. *Can. Dept. Forest., Publ.* **1095**:1–19.

Hough A.F. 1932. Some diameter distributions in forest stands of northwestern Pennsylvania. *J. For.* **30**:933–943.

Hough A.F. and R.D.Forbes, 1943. The ecology and silvics of forests in the high plateaus of Pennsylvania. *Ecol. Mon.* **13**:299–320.

Howe H.F. and R.B.Primack, 1975. Differential seed dispersal by birds of the tree *Casearia nitida* (Flacourtiaceae). *Biotropica* **7**:278–283.

Hozumi K. 1977. Ecological and mathematical considerations on self-thinning in even-aged pure stands. I. Mean plant weight-density trajectory during the course of self-thinning. *Bot. Mag. Tokyo* **90**:165–179.

Hozumi K. and K.Shinozaki, 1970. Studies on the frequency distribution of the weight of individual trees in a forest stand. II. Exponential distribution. *Jap. J. Ecol.* **20**:1–9.

Hozumi, K., K.Shinozaki, and Y.Tadaki, 1968. Studies on the frequency distribution of the weight of individual trees in a forest stand. I. A new approach towards the analysis of the distribution function and the $-\frac{3}{2}$th power distribution. *Jap. J. Ecol.* **18**:10–20.

Hubby J.L. and R.C.Lewontin, 1966. A molecular approach to the study of genic heterozygosity in natural populations. I. The number of alleles at different loci in *Drosophila pseudoobscura. Genetics.* **54**:577–594.

Humphrey R.R. 1974. Fire in the deserts and desert grassland of North America. In Kozlowski T.T. and C.E.Ahlgren (eds.), *Fire and Ecosystems*, pp. 366–400. New York, Academic Press.

Huskins C.L. 1931. The origin of *Spartina townsendii. Genetica* **12**:531–538.

Hutchings M.J. and J.P.Barkham, 1976. An investigation of shoot interactions in *Mercurialis perennis* L., a rhizomatous perennial herb. *J. Ecol.* **64**:723–743.

Hutchison C.B. 1922. Heritable variations in maize. *J. Amer. Soc. Agron.* **14**:73–78.

Hutchinson I. 1976. *Ecological modelling and the stand dynamics of* Pinus caribaea *in Mountain Pine Ridge, Belize*. Ph.D. Thesis, Simon Fraser University, 352 pp.

Hyder D.N. 1972. Defoliation in relation to vegetative growth. In Youngner V.B. and C.M.McKell (eds.), *The Biology and Utilization of Grasses*, pp. 304–317. New York, Academic Press.

Jain S.K. 1975. Population structure and the effects of breeding system. In Frankel O.H. and J.G.Hawkes (eds.), *Crop Genetic Resources for Today and Tomorrow*, pp. 15–36. Cambridge, Cambridge University Press.

Jain S.K. 1976. The evolution of inbreeding in plants. *Ann. Rev. Ecol. Syst.* **7**:469–495.

Jain S.K. and D.R.Marshall, 1967. Genetic changes in a barley population analysed in terms of some life cycle components of selection. *Genetica* **38**:355–374.

Jain S.K. and K.N.Rai, 1974. Population biology of *Avena.* IV. Polymorphism in small populations of *A. fatua. Theor. Appl. Genet.* **44**:7–11.

Jalloq M.C. 1975. The invasion of mole hills by weeds as a possible factor in the degeneration of reseeded pasture. I. The buried viable seed population of mole hills from four reseeded pastures in West Wales. *J. Appl. Ecol.* **12**:643–657.

Janzen D.H. 1966. Coevolution of mutualism between ants and acacias in Central America. *Evolution* **20**:249–275.

Janzen D.H. 1967. Synchronization of sexual reproduction of trees within the dry season in Central America. *Evolution* **21**:620–637.

Janzen D.H. 1969. Seed-eaters versus seed size, number, toxicity and dispersal. *Evolution* **23**:1–27.

Janzen D.H. 1970. Herbivores and the number of tree species in tropical forests. *Amer. Nat.* **104**:501–528.

Janzen D.H. 1971a. Seed predation by animals. *Ann. Rev. Ecol. Syst.* **2**:465–492.

Janzen D.H. 1971b. The fate of *Scheelea rostrata* fruits beneath the parent tree: predispersal attack by bruchids. *Principes* **15**:89–101.

Janzen D.H. 1971c. Escape of juvenile *Dioclea megacarpa* (Leguminosae) vines from predators in a deciduous forest. *Amer. Nat.* **105**:97–112.

Janzen D.H. 1971d. Euglossine bees as long-distance pollinators of tropical plants. *Science* **171**:203–205.

Janzen D.H. 1974. Tropical black water rivers, animals and mast fruiting by the Dipterocarpaceae. *Biotropica* **6**:69–103.

Janzen D.H. 1975a. Behavior of *Hymenaea courbaril* when its predispersal seed predator is absent. *Science* **189**:145–147.

Janzen D.H. 1975b. Interactions of seeds and their insect predators/parasitoids in a tropical deciduous forest. In P.W. Price (ed.), *Evolutionary Strategies of Parasitic Insects and Mites*, pp. 154–186. New York, Plenum Press.

Janzen D.H. 1976a. Reduction of *Mucuna andreana* (Leguminosae) seedling fitness by artificial seed damage. *Ecology* **57**:826–828.

Janzen D.H. 1976b. Why bamboos wait so long to flower. *Ann. Rev. Ecol. Syst.* **7**:347–391.

Janzen D.H. 1977a. Seeding patterns of tropical trees. In Tomlinson P.B. and M.H.Zimmermann (eds.), *Tropical Trees as Living Systems*. Cambridge, Cambridge University Press.

Janzen D.H. 1977b. Variation in seed size within a crop of a Costa Rican *Mucuna andreana* (Leguminosae). *Amer. J. Bot.* **64**:347–349.

Janzen D.H. 1977c. Inter- and intra-crop variation in seed weight of Costa Rican *Ateleia herbert-smithii* (Leguminosae). *Oecologia* (in press).

Janzen D.H. 1977d. What are dandelions and aphids? *Amer. Nat.* **111**:586–589.

Janzen D.H., G.A.Miller, J.Hackforth-Jones, C.M.Pond, K.Hooper and D.P.Janos, 1976. Two Costa Rican bat-generated seed shadows of *Andira inermis* (Leguminoseae). *Ecol.* **57**:1068–1075.

Jeffers J.N.R. 1956. The yield of hazel coppice. *For. Comm. Bull.* **27**:12–18.

Jenkin T.S. 1931a. Self-fertility in perennial rye-grass (*Lolium perenne* L.). *Univ. Coll. Wales. Welsh Plant Breeding Sta. rep.* **12**:100–119.

Jenkin T.S. 1931b. Self-fertility in Italian rye-grass (*Lolium perenne* var. *multiflorum*). *Univ. Coll. Wales. Welsh Plant Breeding Sta. rep.* **12**:120.

Jenkin T.S. 1931c. Fertility in plants of the genus *Phleum*. *Univ. Coll. Wales. Welsh Plant Breeding Sta. rep.* **12**:160.

Jenkins M.T. 1924. Heritable characters in maize. XX. Iojab-stripping, a chlorophyll defect. *J. Heredity.* **15**:467–472.

Jensen H.A. 1969. Content of buried seeds in arable soil in Denmark and its relation to the weed population. *Dansk. Bot. Ark.* **27**:1–56.

Jensen N.F. 1965. Multiline superiority in cereals. *Crop Sci.* **5**:566–568.

Jepson W.L. 1939. *A Flora of California*, Vol. 3, Pt. 1. Berkeley, California: Associated Students Bookstore, University of California.

Jewiss O.R. 1972. Tillering in grasses: its significance and control. *J. Brit. Grassl. Soc.* **27**:65–82.

Johnson E.A. 1975. Buried seed populations in the subarctic forest east of Great Slave Lake, Northwest Territories. *Can. J. Bot.* **53**:2933–2941.

Johnson G.B. 1976a. Genetic polymorphism and enzyme function. In Ayala F. (ed.), *Molecular Evolution*, pp. 46–59. Sunderland, Mass.: Sinaver Associates.

Johnson G.B. 1976b. Hidden alleles at the alpha-glycerophosphate dehydrogenase locus in *Colias* butterflies. *Genetics* **83**:149–167.

Johnson G.B. 1979. Enzyme polymorphism and the physiological phenotype. In Solbrig O.T., S.Jain, G.Johnson and P.H.Raven (eds.), *Topics in Plant Population Biology*. New York, Columbia University Press.

Jones E.W. 1945. The structure and reproduction of the virgin forest of the north temperate zone. *New Phytol.* **44**:130–148.

Juliano J.B. 1940. Viability of some Philippine weed seeds. *Philipp. Agric.* **29**:313–326.

Jurado-Tovar A. and W.A.Compton, 1974. Intergenotypic competition studies in corn (*Zea mays* L.) I. Among experimental hybrids. *Theor. Appl. Genet.* **45**:205–210.

Kadambi K. 1949. On the ecology and silviculture of *Dendrocalamus strictus* in the bamboo forests of Bhadravati Division, Mysore State, and comparative notes on the species *Bambusa arundinacea, Oxythenanthera monostigma* and *Oxythenanthera stocksii. Indian For.* **75**:289–299, 334–349, 398–426.

Karlin S. and E.Nevo (eds.), 1976. *Population Genetics and Ecology*. New York, Academic Press.

Karssen C.M. 1970a. The light promoted germination of *Chenopodium album* L. III. The effect of the photoperiod during growth and development of the plants on the dormancy of the produced seeds. *Acta Bot. Neerl.* **19**:81–94.

Karssen C.M. 1970b. The light promoted germination of *Chenopodium album* L. VI. Pfr requirements during different stages of the germination process. *Acta Bot. Neerl.* **19**:297–312.

Karssen C.M. 1976a. Uptake and effect of absisic acid during induction and progress of radicle growth in seeds of *Chenopodium album. Physiol. Plant.* **36**:259–263.

Karssen C.M. 1976b. Two sites of hormonal action during germination of *Chenopodium album* seeds. *Physiol. Plant.* **36**:264–270.

Kaufman M.L. and A.D.McFadden, 1960. The competitive interaction between barley plants grown from large and small seeds. *Can. J. Pl. Sci.* **40**:623–629.

Kawano S. 1975. The productive and reproductive biology of flowering plants. II. The concept of life history strategy in plants. *J. Coll. Liberal Arts, Toyama Univ.* **8**:51–86.

Kawano S., M.Ihara and M.Suzuki, 1968. Biosystematic studies on *Maianthemum* (Liliaceae-Polygonatae). II. Geography and ecological life history. *Jap. Jour. Bot.* **20**:35–65.

Kawano S. and Y.Nagai, 1975. The productive and reproductive biology of flowering plants. I. Life history strategies of three *Allium* species in Japan. *Bot. Mag. Tokyo* **88**:281–318.

Kawano S., M.Suzuki and S.Kojima, 1971. Biosystematic studies on *Maianthemum* (Liliaceae-Polygonatae). V. Variation in gross morphology, karyology and ecology of North American populations of *M. dilatatum* sensu lato. *Bot. Mag. Tokyo* **84**:299–318.

Kays S. and J.L.Harper, 1974. The regulation of plant and tiller density in a grass sward. *J. Ecol.* **62**:97–105.

Keay R.W.J. 1960. Seeds in forest soils. *Niger. For. Inf. Bull.* (n.s.) **4**:1–12.

Kellman M.C. 1974a. The viable weed seed

content of some tropical agricultural soils. *J. Appl. Ecol.* **11**:669–677.

Kellman M. 1974b. Preliminary seed budgets for two plant communities in coastal British Columbia. *J. Biogeography* **1**:123–133.

Kemp G.A. and L.B.Keith, 1970. Dynamics and regulation of red squirrel *(Tamiasciurus hudsonicus)* populations. *Ecology* **51**:763–779.

Kerner A. 1896. *The Natural History of Plants.* Vol. 2. London, Gresham Publishing Co.

Kern J.J. and R.E.Atkins, 1970. Competition among adjacent rows of grain sorghum of different height genotype. *Agron. J.* **62**:83–86.

Kershaw K.A. 1973. *Quantitative and Dynamic Ecology.* London, Arnold.

Kershaw K.A. 1962. Quantitative ecological studies from Landmannahellir, Iceland. I. *Eriophorum angustifolium. J. Ecol.* **50**: 171–179.

Kevan P.G. 1972. Insect pollination of high Arctic flowers. *J. Ecol.* **60**:831–847.

Khalifa M.A. and C.O.Qualset, 1974. Intergenotypic competition between tall and dwarf wheats. I. Mechanical mixtures. *Crop. Sci.* **14**:795–799.

Khan A.A. 1975. Primary, preventive and permissive roles of hormones in plant systems. *Bot. Rev.* **41**:391–420.

Khan M.A., P.D.Putwain and A.D.Bradshaw, 1975. Population interrelationships. 2. Frequency-dependent fitness in *Linum. Heredity* **34**:145–163.

Khil'mi G.F. 1957. *Theoretical Forest Biogeophysics.* Moscow. (In Russian.)

King J.C. 1955. DDT resistance and the integration of the gene pool. *Amer. Nat.* **89**:39–46.

King T.S. 1975. Inhibition of seed germination under leaf canopies in *Arenaria serpyllifolia, Veronica arvensis* and *Cerastium holosteoides. New Phytol.* **75**:87–90.

Kira T., H.Ogawa and K.Sakazaki, 1953. Intraspecific competition among higher plants. I. Competition-density-yield interrelationships in regularly dispersed populations. *J. Inst. Polytech. Osaka Cy. Univ. D.* **4**:1–16.

Kirchner O. 1905. Über die Wirkung der Selbstbestaubung bei den Papilionaceen. *Naturw. Z. Land-u. Forstw.* **3**:97.

Kirk L.E. 1925. Artificial self-pollination of red clover. *Sci. Agriculture* **5**:179–189.

Kivilaan A. and R.S.Bandurski, 1973. The ninety-year period for Dr. Beal's seed viability experiment. *Am. J. Bot.* **60**:140–145.

Kjoller A. and S.Ødum, 1971. Evidence for longevity of seeds and micro-organisms in permafrost. *Arctic* **24**:230–233.

Knight R. 1960. The growth of *Dactylis glomerata* L. under spaced plant and sward conditions. *Aust. J. Agric. Res.* **11**:451–472.

Knuth P. 1906. *Handbook of Flower Pollination.* Oxford, Clarendon Press.

Koch A.L. 1966. The logarithm in biology. I. Mechanisms generating the log-normal distribution exactly. *J. Theoret. Biol.* **12**:276–290.

Kochumen K.M. 1961. Precious flowering in Dipterocarpaceae. *Malay. Forester* **24**:236.

Koller D. 1972. Environmental control of seed germination. In Kozlowski T.T. (ed.), *Seed Biology,* V. II, pp. 2–101. New York, Academic Press.

Komarek E.V. 1965. Fire ecology—grasslands and man. *Proc. 4th Ann. Tall Timbers Fire Ecology Conf.* 169–220.

Kornicke F. 1890. Ueber autogenetische und heterogenetische Befruchtung bei den Pflanzen. *Verhandl. Naturhis. Vereines Rheinl. u. Westf.* **47**.

Korstian C.F. and W.D.Brush, 1931. Southern white cedar. *USDA Tech. Bull.* **251**.

Koyama H. and T.Kira, 1956. Intraspecific competition among higher plants. VIII. Frequency distribution of individual plant weight as affected by the interaction between plants. *J. Inst. Polytechn. Osaka City Univ. Ser. D* **7**:73–94.

Kozlowski T.T. 1949. Light and water in relation to growth and competition of Piedmont forest tree species. *Ecol. Monogr.* **19**:207–231.

Krefting L.W. and E.I.Roe, 1949. The role of some birds and mammals in seed generation. *Ecol. Monogr.* **19**:269–286.

Kropac A. 1966. Estimation of weed seeds in arable soil. *Pedobiologia* **6**:105–128.

Lambert D.A. 1964. The influence of density and nitrogen in seed production stands of S48 timothy and S215 meadow fescue. *J. agric. Sci.* **63**:35–42.

Landon F.H. 1955. Malayan tropical rain forest. *Malay. Forester* **18**:30–38.

Lang A.L., J.W.Rendleton, and G.H.Dungan, 1956. Influence of population and nitrogen levels on yield and protein and oil content of nine corn hybrids. *Agron. J.* **48**:284–289.

Langer R.H.M. 1956. Growth and nutrition of timothy *(Phleum pratense).* The life history of individual tillers. *Ann. Appl. Biol.* **44**:166–187.

Langer R.H.M. 1957. The effect of time of

cutting on ear production and seed yield in S48 timothy. *J. Brit. Grassl. Soc.* **12**:97–102.

Langer R.H.M. 1963. Tillering in herbage grasses. *Herb. Abst.* **33**:141–148.

Langer R.H.M., S.M.Ryle and O.R.Jewiss, 1964. The changing plant and tiller populations of timothy and meadow fescue swards. 1. Plant survival and the pattern of tillering. *J. Appl. Ecol.* **1**:197–208.

Langridge J. 1962. A genetic and molecular basis for heterosis in *Arabidopsis* and *Drosophila*. *Amer. Nat.* **96**:5–27.

Laude H.M. and E.H.Stanford, 1960. Environmentally induced changes in gene frequency in a synthetic forage variety grown outside the region of adaptation. *Proc. VIII Int. Grassl. Congr.* 180–185.

Law R., A.D.Bradshaw and P.D.Putwain, 1977. Life-history variation in *Poa annua*. *Evolution* **31**:233–246.

Lawson H.M. and J.S.Wiseman, 1978. The effect of weeds on the growth and development of narcissus. *J. Appl. Ecol.* **15**:257–272.

Lazenby A. and H.H.Roger, 1964. Selection criteria in grass breeding. II. Effect, on *Lolium perenne*, of differences in population density, variety and available moisture. *J. Agric. Sci.* **62**:285–298.

Leak W.B. 1964. An expression of diameter distribution for unbalanced, uneven-aged stands and forests. *For. Sci.* **10**:39–50.

Leak W.B. 1975. Age distribution in virgin red spruce and northern hardwoods. *Ecology* **50**:1451–1454.

Lee Y. 1971. Predicting mortality for even-aged stands of lodgepole pine. *For. Chron.* **47**:29–32.

Legay J.M. 1971. Contribution à l'étude de la forme des plantes: discussion d'un modèle de ramification. *Bull. Math. Biophys.* **33**:387–401.

LeHoverou H.N. 1974. Fire and vegetation in the Mediterranean Basin. *Proc. 13th Ann. Tall Timbers Fire Ecol. Conf.* 237–277.

Leigh E.G. 1975. Population fluctuations; community stability and environmental variability. In Cody M.L. and J.Diamond (eds.), *Ecology and Evolution of Communities*, pp. 51–73. Cambridge, Belknap Press of Harvard University.

Lems K. 1960. Botanical notes on the Canary Islands. II. The evolution of plant forms in the islands: *Aeonium*. *Ecology* **41**:1–17.

Leopold L.B. 1971. Trees and streams: the efficiency of branching patterns. *J. Theor. Biol.* **31**:339–354.

Lerner I.M. 1954. *Genetic Homeostasis*. Edinburgh, Oliver and Boyd.

Levin D.A. 1970. Hybridization and evolution—a discussion. *Taxon* **19**:167–176.

Levin D.A. 1972. Competition for pollinator service: a stimulus for the evolution of autogamy. *Evolution* **26**:668–674.

Levin D.A. 1975. Pest pressure and recombination systems in plants. *Amer. Nat.* **109**:437–451.

Levin D.A. and W.W.Anderson, 1970. Competition for pollinators between simultaneously flowering species. *Amer. Nat.* **104**:455–467.

Levin D.A. and H.W.Kerster, 1969. Density-dependent gene dispersal in *Liatris*. *Amer. Nat.* **103**:61–74.

Levin D.A. and H.W.Kerster, 1974. Gene flow in seed plants. *Evolutionary Biology* **7**:139–220.

Levins R. 1968. *Evolution in Changing Environments*. Princeton, Princeton University Press.

Levins R. 1969. Dormancy as an adaptive strategy. *Dormancy and Survival, Soc. Ex. Biol. Symp.* **23**:1–10.

Levins R. 1970. Extinction. *Some Mathematical Questions in Biology* **2**:77–107.

Levins R. 1974. Genetics and Hunger. *Genetics* **78**:67–76.

Lewontine R.C. 1974. *The Genetic Basis of Evolutionary Change*. New York. Columbia University Press.

Lewontin R.C. and L.C.Dunn, 1960. The evolutionary dynamics of a polymorphism in the house mouse. *Genetics* **45**:705–722.

Lewis H. 1962. Catastrophic selection as a factor in evolution. *Evolution* **16**:257–271.

Lewis H. 1973. The origin of diploid neospecies in *Clarkia*. *Amer. Nat.* **107**:161–170.

Lewis H. and P.H.Raven, 1958. Rapid evolution in *Clarkia*. *Evolution* **12**:319–336.

Lewis J. 1973. Longevity of crop and weed seed: survival after 20 years in the soil. *Weed Res.* **13**:179–191.

Liew T.C. and F.O.Wong, 1973. Density, recruitment, mortality and growth of dipterocarp seedlings in virgin and logged-over forests in Sabah. *Malay. Forester* **36**:3–15.

Linhart Y.B. 1976. Density-dependent seed germination strategies in colonizing versus non-colonizing plant species. *J. Ecol.* **64**:375–380.

Linsley E.G. 1966. Pollinating insects of the Galapagos Islands. In Bowman R. (ed.), *The Galapagos*, pp. 225–232. Berkeley, University of California Press.

Lippert R.D. and H.H.Hopkins, 1950. Study of viable seeds in various habitats in mixed prairies. *Trans. Kansas Acad. Sci.* **53**:355–364.

Livingstone R.B. and M.Allessio, 1968. Buried viable seed in successional field and forest stands, Harvard Forest, Mass. *Bull. Torrey Bot. Club* **95**:58–69.

Liubarskii E.L. 1967. The ecology of vegetative reproduction. University of Kacasan. 180 pp. (In Russian.)

Lloyd D.G. 1975a. Breeding systems in *Cotula* IV. Reversion from dioecy to monoecy. *New Phytol.* **74**:125–145.

Lloyd D.G. 1975b. The maintenance of gynodioecy and androdioecy in angiosperms. *Genetica* **45**:325–339.

Lloyd D.G. 1979. Some selective forces affecting the frequency of self-fertilization in angiosperms. *Amer. Nat.* **113**:67–69.

Lloyd D.G. and D.S.Horning, 1979. Distribution of sex in *Coprosma pumila* on Macquarie Island, Australia. *N.Z. J. Bot.* (In press).

Lloyd R.M. 1974. Mating systems and genetic load in pioneer and non-pioneer Hawaiian Pteridophyta. *Bot. J. Linn. Soc.* **69**:23–35.

Löhr E. 1965. Eibe mit 9 Zweigordnungen. *Allg. Forst und Jagdztg.* **136**:267–268.

Long A.G. 1966. Some lower carboniferous fructifications from Berwickshire together with a theoretical account of the evolution of ovules, cupules and carpels. *Trans. Roy. Soc. Edinburgh* **66**:345–375.

Lorimer C.G. 1977. The presettlement forest and natural disturbance cycle in northeastern Maine. *Ecol.* **58**:139–148.

Lutz H.J. 1956. The ecological effects of forest fires in the interior of Alaska. *USDA Tech. Bull.* **1133**.

Lupton F.G.H., A.M.Ali Mohamed and S.Subramaniam, 1967. Varietal differences in growth parameters of wheat and their importance in determining yield. *J. Agr. Sci. Camb.* **69**:111–123.

McArdle R.E. and W.H.Meyer. 1930. The yield of Douglas fir in the Pacific Northwest. *USDA Tech. Bull.* **201**.

MacArthur R.H. 1962. Some generalized theorems of natural selection. *Proc. Nat. Acad. Sci. U.S.* **48**:1893–1897.

MacArthur R.H. 1972. *Geographical Ecology*. New York, Harper and Row.

MacArthur, R.H. and E.O.Wilson, 1967. *The Theory of Island Biogeography*. Princeton, Princeton University Press.

McClure H.E. 1966. Flowering, fruiting and animals in the canopy of a tropical rain forest. *Malay. Forester* **29**:182–203.

McCree K.S. 1970. An equation for the rate of respiration of white clover plants grown under controlled conditions. In Setlik I. (ed.), *Prediction and Measurement of Photosynthetic Productivity.* 221–229 Waageningen, PUDOC.

MacDonald M.A. and P.B.Cavers, 1974. Cauline rosettes—an asexual means of reproduction and dispersal occurring after seed formation in *Barbarea vulgaris* (yellow rocket). *Can. J. Bot.* **52**:913–918.

Mack, R.N. 1976. Survivorship of *Cerastium atrovirens* at Aberffraw, Anglesey. *J. Ecol.* **64**:309–312.

McKell C.M. 1972. Seedling vigour and seedling establishment. In *The Biology and Utilization of Grasses*, pp. 74–89. New York, Academic Press.

MacMahon T.A. and R.E.Kronauer, 1976. Tree structures: deducing the principle of mechanical design. *J. Theor. Biol.* **59**:443–466.

MacMillan, C. 1959. The role of ecotypic variation in the distribution of the central grassland of North America. *Ecol. Monogr.* **29**:287–308.

McNaughton S.J. 1975. *r*- and *K*-selection in *Typha. Amer. Nat.* **109**:251–261.

McWilliam J.R. and B.Griffing, 1965. Temperature-dependent heterosis in maize. *Aust. J. Biol. Sci.* **18**:569–583.

Major J. and W.T.Pyott, 1966. Buried viable seeds in California bunchgrass sites and their bearing on the definition of a flora. *Vegetatio Acta Geobotanica* **13**:253–282.

Markert C.L. 1968. The molecular basis for isozymes. *Ann. N.Y. Acad. Sci.* **151**:14–40.

Marks P.L. 1974. The role of pin cherry (*Prunus pennsylvanica* L.) in the maintenance of stability in northern hardwood ecosystems. *Ecol. Mon.* **44**:73–88.

Mar-Möller C., D.Müller and J.Nielsen, 1954. Loss of branches in European beech. *Det forstlige Forsøg. Danmark* **21**:253–271.

Mather K. 1940. Outbreeding and separation of the sexes. *Nature* **145**:484–486.

Mather K. 1941. Variation and selection of polygenic characters. *Jour. Genet.* **41**:159–193.

Mather K. 1943. Polygenic inheritance and natural selection. *Biol. Rev.* **18**:32–64.

Matthews C.P. and D.F.Westlake, 1969. Estimation of production by a population of higher plants subject to high mortality. *Oikos* **20**:156–160.

Mauve K. 1931. Ueber Bestandesaufbau, Zuwachs — verhältnisse und verjüngung im galizischen Karpathen Urwald. *Mitt. Forstwirt. Forstwissen.* **2**:257–311.

Mayer A.M. and A.Poljakoff-Mayber, 1975. *The Germination of Seeds.* New York, Pergamon Press.

Maynard Smith J. 1971a. What use is sex? *J. Theor. Biol.* **30**:319–335.

Maynard Smith J. 1971b. The origin and maintenance of sex. In Williams G.C. (ed.), *Group Selection,* pp. 163–175. Chicago, Aldine-Atherton.

Mayr E. 1963. *Animal Species and Evolution.* Cambridge, Mass., Harvard University Press.

Mayr E. 1973. The recent historiography of genetics. *Jour. Hist. of Biology* **6**:125–154.

Mayr E. 1976. The recent historiography of Genetics. In Mayr E. *Evolution and the Diversity of Life,* pp. 329–353. Cambridge, Mass., Belknap-Press of Harvard University Press.

Mayr E. 1976. *Evolution and the Diversity of Life.* Cambridge, Mass., Harvard University Press.

Mead R. 1968. Measurement of competition between individual plants in a population. *J. Ecol.* **56**:35–45.

Medina E. 1971. Effect of nitrogen supply and light intensity during growth on the photosynthetic capacity and carboxydismutase activity of leaves of *Atriplex patula* ssp. *hastata. Carnegie Inst. Yearbook* **70**:551–559.

Medway L. 1972. Phenology of a tropical rain forest in Malaya. *Biol. J. Linn. Soc., Lond.* **4**:117–146.

de la Mensbruge G. 1966. Germination and seedlings of tree species of the tropical moist forest of the Ivory Coast. *Publ. Centre Tech. For. Trop.* **26**:389 pp.

Meusel H. and G.Mörchen, 1977. Zur ökogeographischen und morphologischen Differenzierung einiger *Scrophularia* — Arten. *Flora* **166**:1–20.

Meyer H.A. 1952. Structure, growth and drain in balanced uneven-aged forests. *J. For.* **50**:85–92.

Meyer W.H. 1929. Yields of second-growth spruce and fir in the northeast. *USDA Tech. Bull.* **142**.

Meyer W.H. 1937. Yield of even-aged stands of sitka spruce and western hemlock. *USDA Tech. Bull.* **544**.

Meyer W.H. 1938. Yield of even-aged stands of ponderosa pine. *USDA Tech. Bull.* **630**.

Miles J. 1972. Experimental establishment of seedlings on a southern English heath. *J. Ecol.* **60**:225–234.

Miles J. 1973. Early mortality and survival of self-sown seedlings in Glenfeshie, Invernesshire. *J. Ecol.* **61**:93–98.

Miller R.B. 1923. First report on a forestry survey in Illinois. *Illinois Nat. Hist. Survey Bull.* **14**:291–377.

Millington W.F. and W.R.Chaney, 1973. Shedding of shoots and branches. In Kozlowski T.T. (ed.), *Shedding of Plant Parts,* pp. 149–203. New York, Academic Press.

Milton W.E.J. 1936. The buried viable seeds of enclosed and unenclosed hill land. *Welsh Pl. Breed. Stn. Bull.* Series **H14**:58–84.

Miyaji K.-I. and H.Tagawa, 1973. A life table of the leaves of *Tilia japonica* Simonkai. *Rep. Ebino Biol. Lab. Kyushu Univ.* **1**:98–108. (Japanese, English summary and legends.)

Moll R.H., J.H.Lonnquist, J.Velez Fortuno and E.L.Johnson, 1965. The relationship of heterosis and genetic divergence in maize. *Genetics* **52**:139–144.

Monyo J.H., A.D.R.Ker and M.Campbell, 1976. *Intercropping in Semi-arid Areas.* International Development Centre, Ottawa.

Mooney H.A. 1976. Some contributions of plant physiological ecology to plant population biology. *Syst. Bot.* **1**:269–283.

Mooney H.A. and W.D.Billings, 1961. Comparative physiological ecology of arctic and alpine populations of *Oxyria digyna. Ecol. Monogr.* **31**:1–29.

Mooney H.A. and S.L.Gulmon. 1979. Environmental and evolutionary constraints on the photosynthetic characteristics of higher plants. In Solbrig O.T., S.Jain, G.Johnson and P.H.Raven (eds.), *Topics in Population Biology.* New York, Columbia University Press.

Moore W.S. 1976. Components of fitness in the unisexual fish, *Poeciliopsis monacha-occidentalis. Evolution* **30**:564–578.

Morey H.F. 1936. Age–size relationships of Hearts Content, a virgin forest in northwestern Pennsylvania. *Ecology* **17**:251–257.

Mori S.A. and J.A.Kallunki, 1976. Phenology and floral biology of *Gustavia superba* (Lecythidaceae) in Central Panama. *Biotropica* **8**:184–192.

Morley F.H.W., C.I.Davern, V.E.Rogers and I.W.Peak, 1962. Natural selection among strains of *Trifolium subterraneum.* In Leeper G.W. (ed.), *The Evolution of Living Organisms,* pp. 181–190. Melbourne, Melbourne University Press.

Mosquin T. 1966. Reproductive specialization as a factor in the evolution of the Canadian flora. In Taylor P.L. and R.A.Ludwig (eds.), *The Evolution of Canada's Flora,* pp. 43–65. Toronto, University of Toronto Press.

Mosquin T. 1971. Competition for pollinators

as a stimulus for the evolution of flowering time. *Oikos* **22**:398–402.

Mueller-Dombois D. and H.Ellenberg, 1974. *Aims and Methods of Vegetation Ecology.* New York, John Wiley.

Mulcahy D.L. (ed.), 1975. *Gamete Competition in Plants and Animals.* Amsterdam, North-Holland Publishing Co.

Müller H. 1883. *The Fertilization of Flowers.* London, Macmillan.

Mulligan G.A. and J.N.Finlay, 1970. Reproductive systems and colonization in Canadian weeds. *Can. J. Bot.* **48**:859–860.

Münch E. 1938. Untersuchungen über die Harmonie der Baumgestalt. *Jahr. wiss. Bot.* **86**:581–673.

Nakoneshny W. and G.Friesen, 1961. The influence of commercial fertilizer treatment on weed competition in spring sown wheat. *Can. J. Pl. Sci.* **41**:231–238.

Naveh Z. 1974. Effects of fire in the Mediterranean Region. In Kozlowski T.T. and C.E.Ahlgren (eds.), *Fire and Ecosystems,* pp. 401–434. New York, Academic Press.

Naylor R.E.L. 1972. Aspects of the population dynamics of the weed *Alopecurus myosuroides* Huds. in winter cereal crops. *J. appl. Ecol.* **9**:127–139.

Naylor R.E.L. 1976. Changes in the structure of plant populations. *J. appl. Ecol.* **13**:513–521.

Nei, M. 1975. *Molecular Population Genetics and Evolution.* Amsterdam, North-Holland Publishing Co.

Nei M., T.Maruyama and R.Chakraborty, 1975. The bottleneck effect and genetic variability in populations. *Evolution* **29**:1–10.

Nei M. and K.Syakudo, 1958. The estimation of outcrossing in natural populations. *Jap. J. Genet.* **33**:46–51.

Nelson, J.F. and R.M.Chew, 1977. Factors affecting seed reserves in the soil of a Mojave desert ecosystem, Rock Valley, Nye County, Nevada. *Am. Midl. Nat.* **97**:300–320.

Nelson, O.E. and A.S.Ohlrogge, 1957. Differential responses to population pressures by normal and dwarf lines of maize. *Science* **125**:1200.

Nevling L.I. 1971. The ecology of an elfin forest in Puerto Rico. 16. The flowering cycle and an interpretation of its seasonality. *J. Arnold Arb.* **52**:586–613.

Newton I. 1967. The feeding ecology of the bullfinch (*Pyrrhula pyrrhula* L.) in Southern England. *J. Anim. Ecol.* **36**:721–744.

Ng F.S.P. 1966. Age at first flowering in dipterocarps. *Malay. Forester* **29**:290–295.

Nicholson D.I. 1960. Light requirements of seedlings of five species of Dipterocarpaceae. *Malay. Forester* **23**:344–356.

Nicholson D.I. 1965. A review of natural regeneration in the dipterocarp forests of Sabah. *Malay. Forester* **28**:4–25.

Nielsen E.L. and D.C.Smith, 1959. Chlorophyll inheritance patterns and extent of natural self-pollination in Timothy. *Euphytica* **8**:169–179.

Nieto J.H., M.A.Brondo and J.T.Gonzales, 1968. Critical periods of the crop growth cycle for competition from weeds. *PANS* (c) **14**:159–166.

Nieto J.H. and D.W.Staniforth, 1961. Corn-foxtail competition under various production conditions. *Agron. J.* **53**:1–5.

Nilsson-Ehle H. 1909. Kreuzungsuntersuchungen an Hafer und Weizen. *Lunds Univ. Arsskr* **2**:5(2) 1–122.

Njoku E. 1964. Seasonal periodicity in the growth and development of some forest trees in Nigeria. II. Observations on seedlings. *J. Ecol.* **52**:19–26.

Nozeran R., L.Bancilhon and P.Neville, 1971. Intervention of internal correlations in the morphogenesis of higher plants. *Adv. Morphogen.* **9**:1–66.

Obeid M., D.Machin and J.L.Harper, 1967. Influence of density on plant to plant variation in fiber flax, *Linum usitatissimum. Crop Sci.* **7**:471–473.

Ødum S. 1965. Germination of ancient seeds; floristical observations and experiments with archaeologically dated soil samples. *Danske. Botanisk. Arkiv.* **24**:2.

Ødum S. 1974. Seeds in ruderal soils their longevity and contribution to the flora of disturbed ground in Denmark. *Proc. 12th Brit. Weed Contr. Conf.* pp. 1131–1144.

Ogden J. 1968. *Studies on reproductive strategy with particular reference to selected composites.* Ph.D. thesis, University of Wales.

Ogden J. 1974. The reproductive strategy of higher plants. II. The reproductive strategy of *Tussilago farfara* L. *J. Ecol.* **62**:291–324.

Oldeman R.A.A. 1974. *L'Architecture de la Forêt Guyanaise.* Paris, ORSTOM.

Oliver C.D. 1975. *The development of northern red oak (Quercus rubra L.) in mixed species even-aged stands in central New England.* Ph.D. Thesis, Yale University.

Oliver C.D. 1978. The development of northern red oak in mixed stands in central New

England. *Yale Univ. School of For. Envir. Stud. Bull.* **91:**1–63.

Oliver C.D. and E.P.Stephens, 1977. Reconstruction of a mixed-species forest in central New England. *Ecology* **58:**562–572.

Oohata S. and T.Shidei, 1971. Studies on the branching structure of trees. I. Bifurcation ratio of trees in Horton's law. *Jap. J. Ecol.* **21:**7–14.

Oomes M.J.M. and W.Th.Elberse, 1976. Germination of six grassland herbs in microsites with different water contents. *J. Ecol.* **64:**745–755.

Oosting H.J. and M.E.Humphreys, 1940. Buried viable seeds in a successional series of old field and forest soils. *Bull. Torrey Bot. Club.* **57:**253–273.

Ovington J.D. 1953. A study of invasion by *Holcus mollis* L. *J. Ecol.* **41:**35–52.

Opler P.A., H.G.Baker and G.W.Frankie, 1975. Reproductive biology of some Costa Rican *Cordia* species (Boraginaceae). *Biotropica* **7:**234–247.

Palvakul M., V.C.Finkner and D.L.Davis, 1973. Blendability of phenotypically similar and dissimilar winter barley cultivars. *Agron. J.* **65:**74–77.

Parsons R.A. 1971. Extreme environment heterosis and genetic loads. *Heredity* **26:**479–483.

Paterson J.G., W.J.R.Boyd and N.A.Goodchild, 1976. Effect of temperature and depth of burial on the persistence of seed of *Avena fatua* in Western Australia. *J. appl. Ecol.* **13:**841–847.

Peñalosa J. 1975. *Shoot Dynamics of Tropical Lianas.* Ph.D. Thesis. Harvard University.

Penning de Vries F.W.T. 1975. Use of assimilates in higher plants. In Cooper J.P. (ed.), *Photosynthesis and Productivity in Different Environments,* pp. 459–480. Cambridge, Cambridge University Press.

Perttula U. 1941. Untersuchungen über die generative und vegetative Vermehrung der Blütenpflanzen in der Wald-, Hain-, Wiesen-, und Hainfelsen vegetation. *Ann. Acad. Sci. Fenn. Ser. A. Tom.* **58:**1–388.

Pettitt J. 1970. Heterospory and the origin of the seed habit. *Biol. Rev.* **45:**401–415.

Philip M.S. 1968. The dynamics of seedling populations in moist semi-deciduous tropical forest in Uganda. Aberdeen, Dept. of Forestry.

Piñero D., J.Sarukhán and P.Alberdi (in preparation) Yearly budgets and whole life strategies of energy allocation in the palm *Astrocaryum mexicanum.*

Piñero D., J.Sarukhán and E.González, 1977. Estudios demográficos en plantas. *Astrocaryum mexicanum.* I. Estructura de las poblaciones. *Bol. Soc. Bot. Mex.* **37:**69–118.

Platt W.J. 1976. The natural history of a fugitive prairie plant (*Mirabilis hirsuta* (Pursh.) MacM.). *Oecologia* **22:**399–409.

Pojar J. 1974. Reproductive dynamics of four plant communities of southwestern British Columbia. *Can. J. Bot.* **52:**1219–1234.

Pollard D.F.W. 1971. Mortality and annual changes in distribution of above-ground biomass in an aspen sucker stand. *Can. J. For. Res.* **1:**262–266.

Pollard D.F.W. 1972. Above-ground dry matter production in three stands of trembling aspen. *Can. J. For. Res.* **2:**27–33.

Poore M.D. 1976. Studies in Malaysian rain forest. I. The forest on the Triassic sediment in Jengka Forest Reserve. *J. Ecol.* **56:**143–196.

Popay A.I. and E.H.Roberts, 1970a. Factors involved in the dormancy and germination of *Capsella bursa-pastoris* and *Senecio vulgaris. J. Ecol.* **58:**103–122.

Popay A.I. and E.H.Roberts, 1970b. Ecology of *Capsella bursa-pastoris* and *Senecio vulgaris* in relation to germination behavior. *J. Ecol.* **58:**123–139.

Porsild A.E., C.R.Harington and G.A.Mulligan, 1967. *Lupinus arcticus* Wats. grown from seed of Pleistocene age. *Science* **158:**113–114.

Porter D.M. 1976. Geography and dispersal of Galapagos Islands vascular plants. *Nature* **264:**745–746.

Prout T. 1965. The estimation of fitness from genotypic frequencies. *Evolution* **19:**546–551.

Prout T. 1969. The estimation of fitness from population data. *Genetics* **63:**949–967.

Puckeridge D.W. and C.M.Donald, 1967. Competition among wheat plants sown at a wide range of densities. *Aust. J. Agric. Res.* **18:**193–211.

Putwain P.D. and J.L.Harper, 1970. Studies in the dynamics of plant populations. III. The influence of associated species on populations of *Rumex acetosa* and *R. acetosella* in grassland. *J. Ecol.* **58:**251–264.

Putwain P.D. and J.L.Harper, 1972. Studies in the dynamics of plant populations. V. Mechanisms governing the sex ratio in *Rumex acetosa* and *R. acetosella. J. Ecol.* **60:**113–129.

Qualset C.O. 1968. Population structure and performance in wheat. *Proc. 3rd Int. Wheat Genet. Symp.* pp. 397–402.

Rabinovich J.E. 1975. Demographic strategies in animal populations: a regression analysis. In Golley F.B. and E.Medina (eds.), *Tropical Ecological Systems*, pp. 19–40. Berlin, Springer-Verlag.

Rabotnov T.A. 1969. On coenopopulations of perennial herbaceous plants in natural coenoses. *Vegetatio* **19:**87–95.

Raven P.H. 1979. Future directions in plant population biology. In Solbrig O.T., S.Jain, G.Johnson and P.H.Raven (eds.), *Topics in Plant Population Biology*. New York, Columbia University Press.

Reeves T.G. 1976. Effect of annual ryegrass (*Lolium rigidum* Gand.) on yield of wheat. *Weed Res.* **16:**57–63.

Rehfeldt G.E. and D.T.Lester, 1969. Specialization and flexibility in genetic systems of forest trees. *Silvae Genet.* **18:**118–123.

Reich V.H. and R.E.Atkins, 1970. Yield stability of four population types of grain sorghum in different environments. *Crop Sci.* **10:**511–517.

Resvoll T.R. 1925. *Rubus chamaemorus* Die geogr. Verbrietung d. Pflanze u. ihre Verbreitungsmittel. *Festschrift Carl Schröfer (Zurich)* pp. 224–241.

Resvoll T.R. 1929. *Rubus chamaemorus* L.—A morphological-biological study. *Nyt. Mag. f. Naturv.* **67:**55–129.

Rhodes I. 1970. The production of contrasting genotypes of perennial ryegrass in monocultures and mixed cultures of varying complexity. *J. Br. Grassl. Soc.* **25:**285–288.

Richards P.W. 1952. *The Tropical Rainforest.* Cambridge, Cambridge University Press.

Rick C.M. 1966. Some plant-animal relations in the Galapagos. In Bowman R. (ed.), *The Galapagos*, pp. 215–224. Berkeley, University of California Press.

Ridley H.N. 1930. *Dispersal of Plants Throughout the World.* Ashford, Reeve.

Roberts E.H. 1972a. Storage environment and the control of viability. In Roberts E.H. (ed.), *Viability of Seeds*, pp. 14–58. Syracuse, Syracuse University Press.

Roberts E.H. 1972b. Cytological genetical and metabolic changes associated with loss of viability. In Roberts E.H. (ed.), *Viability of Seeds*, pp. 253–306. Syracuse, Syracuse University Press.

Roberts E.H. 1972c. Dormancy: a factor affecting seed survival in soil. In Roberts E.H. (ed.), *Viability of Seeds*, pp. 321–359. Syracuse, Syracuse University Press.

Roberts H.A. 1968. The changing population of viable weed seeds in an arable soil. *Weed Res.* **8:**253–256.

Roberts H.A. 1970. Viable weed seeds in cultivated soils. *Rep. Nat. Veg. Res. Stn.* **1969:**25–38.

Roberts H.A. 1976. Weed competition in vegetable crops. *Ann Appl. Biol.* **83:**521–527.

Roberts H.A. and P.M.Feast, 1972. Emergence and longevity of seeds of annual weeds in cultivated and undisturbed soil. *J. Appl. Ecol.* **10:**133–143.

Roberts H.A. and P.M.Feast, 1973. Changes in the numbers of viable weed seeds in soil under different regimes. *Weed Res.* **13:**298–303.

Rollins R.C. 1963. The evolution and systematics of *Leavenworthia* (Cruciferae). *Contrib. Gray Herb. (Harvard Univ.)* **192:**3–98.

Rollins R.C. and O.T.Solbrig, 1973. Interspecific hybridization in *Lesquerella. Contrib. Gray Herb. (Harvard Univ.)* **203:**3–48.

Roos F.H. and J.A.Quinn, 1977. Phenology and reproductive allocation in *Andropogon scoparius* (Graminae) populations in communities of different successional stages. *Amer. J. Bot.* **64:**535–540.

Roose M.L. and L.D.Gottlieb, 1976. Genetic and biochemical consequences of polyploidy in *Tragopogon. Evolution* **30:**818–830.

Rossiter R.C. 1974. The relative success of strains of *Trifolium subterraneum* L. in binary mixtures under field conditions. *Aust. J. Agric. Res.* **25:**757–766.

Rosenzweig M.L. and P.W.Sterner, 1970. Population ecology of desert rodent communities: body size and seed husking as bases for heteromyid coexistence. *Ecology* **51:**217–224.

Rowe J.S. 1955. Factors influencing white pine reproduction in Manitoba and Saskatchewan. *Can., Dept. North. Affairs Natur. Resour., Forest Res. Div., Tech. Note 3.*

Ruiz Zapata T.R. and M.T.Kalin de Arroyo, 1978. Plant reproductive ecology of species of a secondary deciduous tropical forest in Venezuela. *Biotropica.* **10:**221–230.

Russell W.A. 1969. Hybrid performance of maize inbred lines selected by test cross performance in low and high plant densities. *Crop Sci.* **9:**185–188.

Ryle G.J.A. 1964. The influence of date of origin of the shoot and level of nitrogen on ear size in three perennial grasses. *Ann. Appl. Biol.* **53:**511–523.

Sagar G.R. 1970. Factors controlling the size of plant populations. *Proc. 10th Brit. Weed Control Conf.* 3:965–979.

Sagar G.R. and J.L.Harper, 1960. Factors affecting the germination and early establishment of plantains *(Plantago lanceolata, P. media* and *P. major)*. In Harper J.L. (ed.), *The Biology of Weeds*, pp. 236–245. *Br. Ecol. Soc. Symp.* No. 1.

Sagar G.R. and A.M.Mortimer, 1976. An approach to the study of the population dynamics of plants with special reference to weeds. *Ann. Appl. Biol.* 1:1–47.

Salisbury E.J. 1942. *The Reproductive Capacity of Plants.* London, Bell.

Salisbury E.J. 1961. *Weeds and Aliens.* London, Collins Press.

Salisbury E.J. 1974. Seed size, and mass in relation to environment. *Proc. Royal Soc. Lond. B* 186:83–88.

Salisbury E.J. 1975. The survival value of modes of dispersal. *Proc. Royal Soc. Lond. B* 188:183–188.

Sanchez J. and F.Davis, 1969. Growth inhibitors in Kikuyu *(Pennisetum clandestinum)* as factors of its competitive ability. *I. Semm. Soc. Colomb. de Contr. des Malezas y Fisiol. Vegetal (COMALFI)*, Bogotá 23–24, 58–59 (Abstr.)

Sarkissian, I.V. and H.K.Skrivaslava, 1967. Mitochondrial polymorphism in maize. II. Further evidence of correlation of mitochondrial complementation and heterosis. *Genetics* 57:843–850.

Sarukhán J. 1974. Studies on plant demography: *Ranunculus repens* L., *R. bulbosus* L., and *R. acris* L. II. Reproductive strategies and seed population dynamics. *J. Ecol.* 62:675–716.

Sarukhán J. 1976. On selective pressures and energy allocation in populations of *Ranunculus repens* L., *R. bulbosus* L., and *R. acris* L. *Ann. Missouri Bot. Gard.* 63:290–308.

Sarukhán J. 1978. Studies on the demography of tropical trees. In Tomlinson P.B. and M.H.Zimmermann (eds.), *Tropical Trees as Living Systems.* pp. 163–184. Cambridge University Press.

Sarukhán J. and M.Gadgil, 1974. Studies on plant demography: *Ranunculus repens* L., *R. bulbosus* L., and *R. acris* L. III. A mathematical model incorporating multiple modes of reproduction. *J. Ecol.* 62:921–936.

Sarukhán J. and J.L.Harper, 1973. Studies on plant demography: *Ranunculus repens* L., *R. bulbosus* L., and *R. acris* L. I. Population flux and survivorship. *J. Ecol.* 61:675–716.

Sarukhán J. and D.Piñero (in preparation)

Studies on the demography of tropical plants: *Astrocaryum mexicanum* (Palmae) I. Population flux and reproductive schedules.

Saucier J.R., A.Clark III and R.G.McAlpine, 1972. Above ground biomass yields of short-rotation sycamore. *Wood Sci.* 5:1–6.

Sauer J. and G.Struik, 1964. A possible ecological relation between soil disturbance, light flash, and seed germination. *Ecology* 45:884–886.

Savile D.B.O. 1972. Arctic adaptations in plants. *Canada Dept. Agr. Research Branch Monograph* 6:1–81.

Schafer D.E. and D.O.Chilcote, 1969. Factors influencing persistence and depletion in buried seed populations. I. A model for analysis of parameters of buried seed persistence and depletion. *Crop Sci.* 9:417–419.

Schafer D.E. and D.O.Chilcote, 1970. Factors influencing persistence and depletion in buried seed populations. II. The effects of soil temperature and moisture. *Crop Sci.* 10:342–345.

Schaffer W.M. and M.D.Gadgil, 1975. Selection for optimal life histories in plants. In Cody M. and J.Diamond (eds.), *Ecology and Evolution of Communities*, pp. 142–157. Cambridge, Harvard University Press.

Schaffner J.H. 1938. Spreading of *Opuntia* in overgrazed pastures in Kansas. *Ecology* 19:348–350.

Schreiber M.M. 1967. A technique for studying weed competition in forage legume establishment. *Weeds* 15:1–4.

Schumacher F.X. 1926. Yield, stand, and volume tables for white fir in the California pine region. *Univ. Calif. College Agric. Bull.* 407.

Schulz J.R. 1960. *Ecological Studies on the Rain Forest of Northern Suriname, The Vegetation of Suriname*, Vol. 2. Amsterdam, North-Holland.

Schutz W.H. and C.A.Brim, 1971. Inter-genotypic competition in soybeans. III. An evaluation of stability in multiline mixtures. *Crop Sci.* 11:681–689.

Scott R.K. and S.J.Wilcockson, 1976. Weed biology and the growth of sugar beet. *Ann. Appl. Biol.* 83:331–335.

Sedcole J.R. and R.J.Clements, 1973. Studies of genotype and spacing interactions for herbage yield, using a modified diallel analysis. *J. Agric. Sci.* 80:97–104.

Selander R.K. 1976. Genetic variation in natural population. In Ayala F. (ed.), *Molecular Evolution*, pp. 21–45. Sunderland, Mass., Sinauer.

Selman M. 1970. The population dynamics of

Avena fatua in continuous spring barley. Desirable frequency of spraying with triallate. *Proc. 10th Br. Weed Contr. Conf.* 1176–1184.

Sharitz R.R. and J.F.McCormick, 1973. Population dynamics of two competing annual plant species. *Ecology* **54**:723–740.

Shaw M.W. 1968. Factors affecting the natural regeneration of sessile oak *(Quercus petrea)* in North Wales. II. Acorn losses and germination under field conditions. *J. Ecol.* **56**:647–660.

Shirley H.L. 1929. The influence of light intensity and light quality upon the growth of plants. *Amer. J. Bot.* **16**:354–390.

Sheldon J.C. 1974. The behavior of seeds in soil. III. The influence of seed morphology and the behavior of seedlings on the establishment of plants from surface lying seeds. *J. Ecol.* **62**:47–66.

Sheldon J.C. and F.M.Burrows, 1973. The dispersal effectiveness of the achenepappus units of selected Compositae in steady winds with convection. *New Phytol.* **72**:665–675.

Shontz N.N. and J.P.Shontz, 1972. Rapid evolution in populations of *Galinsoga ciliata* in Western Massachusetts. *Amer. Midl. Nat.* **88**:183–199.

Simmonds F.J. 1951. Further effects of the defoliation of *Cordia macrostachya* (Jacq.) R. & S. *Can. Entomol.* **83**:24–27.

Simmonds N.W. 1962. Variability in crop plants, its use and conservation. *Biol. Rev.* **37**:422–465.

Simpson B.B., J.L.Neff and A.R.Moldenke, 1977. *Prosopis* flowers as a resource. In Simpson B.B. (ed.), *Mesquite*, pp. 84–107. Stroudsburg, Penn., Dowden, Hutchinson & Ross.

Simpson D.M. 1954. Natural cross-pollinations in cotton. *USDA Tech. Bull.* **1094**:1–17.

Simpson D.M. and E.N.Duncan, 1956. Cotton pollen dispersal by insects. *Agron. J.* **48**:305–308.

Singh L.B. 1960. *The Mango.* London, Leonard Hill.

Singh R.S. and S.K.Jain, 1971. Population biology of *Avena*. II. Isoenzyme polymorphism in populations of the Mediterranean region and Central California. *Theor. Appl. Genet.* **41**:79–84.

Singh R.S., R.C.Lewontin and A.A.Felton, 1976. Genetic heterogeneity within electrophoretic 'alleles' of xanthine dehydrogenase in *Drosophila pseudoobscura*. *Genetics* **84**:609–629.

Sinnott E.W. 1927. A factorial analysis of certain shape characters in squash fruits. *Amer. Nat.* **61**:333–344.

Sinnott E.W. and D.Hammond, 1930. Factorial balance in the determination of fruit shape in *Cucurbita*. *Amer. Nat.* **64**:509–524.

Smith A.J. 1975. Invasion and ecesis of bird-disseminated woody plants in a temperate forest sere. *Ecol.* **56**:19–34.

Smith A.P. and J.O.Palmer, 1976. Vegetative reproduction and close packing in a successional plant species. *Nature* **261**:232–233.

Smith C.C. 1970. The coevolution in pine squirrels *(Tamiasciurus)* and conifers. *Ecol. Monogr.* **40**:349–371.

Smith C.C. 1975. The coevolution of plants and seed predators. In Gilbert L.E. and P.H.Raven (eds.), *Coevolution of Animals and Plants*, pp. 51–77. Austin, University of Texas Press.

Smith C.C. and S.D.Fretwell, 1974. The optimal balance between size and number of offspring. *Amer. Nat.* **108**:499–506.

Smith H. 1972. Light quality and germination: ecological implications. In Heydecker (ed.), *Seed Ecology*, pp. 219–323. Pennsylvania State University Press. University Park,

Smith H. 1975. *Phytochrome and Photomorphogenesis.* New York, McGraw-Hill.

Smith J.H.G. 1968. Growth and yield of red alder in British Columbia. In Trappe J.M., J.F.Franklin, R.F.Tarrant and G.M.Hansen (eds.), *Biology of Alder*, pp. 273–286. Washington, USDA Forest Service.

Smith R.F. 1970. The vegetation structure of a Puerto Rican rain forest before and after short-term gamma irradiation. In Odum H.T. and R.F.Pigeon (eds.), *A Tropical Rain Forest*, pp. D103–D140. Oak Ridge, Tenn., U.S. Atomic Energy Comm.

Smithers L.A. 1961. Lodgepole pine in Alberta. *Canada Dept. For. Bull.* **127**.

Smythe N. 1970. Relationship between fruiting seasons and seed dispersal methods in a neotropical forest. *Amer. Nat.* **104**:25–35.

Snaydon R.W. 1971. An analysis of competition between plants of *Trifolium repens* collected from contrasting soils. *J. Appl. Ecol.* **8**:687–697.

Snaydon R.W. 1978a. Genetic changes in pasture populations. In Wilson J.R. (ed.), *Plant Relations in Pastures*, pp. 253–269. Melbourne, CSIRO.

Snaydon, R.W. 1978b. Indigenous species in perspective. *Proc. Brit. Crop Protection Conf.* (Weeds) 1978 (in press).

Snaydon, R.W. and M.S.Davies, 1972. Rapid population differentiation in a mosaic environment. II. Morphological variation in *Anthoxanthum odoratum. Evolution* **26**:390–405.

Snaydon R.W. & M.S.Davies, 1976. Rapid population differentiation in a mosaic environment. IV. Populations of *Anthoxanthum odoratum* at sharp boundaries. *Heredity* **37**:9–25.

Snaydon R.W. and J.E.Elston, 1976. Flows, cycles and yields in agricultural ecosystems. In Duckham A.N., J.G.W.Jones and E.H.Roberts (eds.), *Food Production and Consumption,* pp. 43–60. Amsterdam, North Holland Publ.

Sork V. and D.H.Boucher, 1977. Dispersal of sweet pignut hickory in a year of low fruit production, and the influence of predation by a curculionid beetle. *Oecologia* **28**:289–299.

Solbrig O.T. 1972. Breeding system and genetic variation in Leavenworthia. *Evolution* **26**:155–160.

Solbrig O.T. 1976. On the relative advantages of cross and self-fertilization. *Ann. Mo. Bot. Gard.* **63**:262–276.

Solbrig O.T. and P.Cantino, 1975. Reproductive adaptations in *Prosopis* (Leguminosae, Mimosoideae). *Jour. Arnold Arb.* **56**: 185–210.

Solbrig O.T., S.Jain, G.Johnson and P.Raven (eds.), 1979. *Topics in Plant Population Biology*. New York, Columbia University Press.

Solbrig O.T. and R.C.Rollins, 1977. The evolution of autogamy in species of the mustard genus. *Leavenworthia. Evolution* **31**:265–281.

Solbrig O.T. and B.B.Simpson, 1974. Components of regulation of a population of dandelions in Michigan. *J. Ecol.* **62**:473–486.

Solbrig O.T. and B.B.Simpson, 1977. A garden experiment on competition between biotypes of the common dandelion *(Taraxacum officinale). J. Ecol.* **65**:427–430.

Söyrinki N. 1938. Studien über die generative und vegetative Vermehrung der Samenpflanzen in der alpinen Vegetation Petsamo-Lapplands. *Ann. Bot. Soc. Zool.-Bot. Fenn. Vanamo* **11**:1–311.

Spatz G. and D.Mueller-Dombois, 1973. The influence of feral goats on koa tree reproduction in Hawaii Volcanoes National Park. *Ecology* **54**:870–876.

Spiegelman M. 1968. *Introduction to Demography*. Cambridge, Mass., Harvard University Press.

Sporne K.R. 1976. Character correlations among angiosperms and the importance of fossil evidence in assessing their importance. In Beck C.D. (ed.), *Origin and Evolution of Early Angiosperms*. New York, Columbia University Press.

Sprague, G.T. 1969. Germ Plasm Manipulations of the Future. In Eastin, J.D., F.A. Haskins, C.Y.Sullivan and C.H.M.Van Bavel (eds.), *Physiological Aspects of Crop Yield,* pp. 375–387. Madison, Wiss. Am. Soc. of Agronomy.

Sprague G.F. and J.F.Schuler, 1961. The frequencies of seed and seedling abnormalities in maize. *Genetics* **46**:1713–1720.

Spring P.E., M.L.Brewer, J.R.Brown and M.E.Fanning, 1974. Population ecology of loblolly pine *Pinus taeda* in an old field community. *Oikos* **25**:1–6.

Spurr S.H., L.J.Young, B.V.Barnes and E.L.Hughes 1957. Nine successive thinnings in a Michigan white pine plantation. *J. For.* **55**:7–13.

Stauffer R.C. 1975. *Charles Darwin's Natural Selection*. Cambridge, Cambridge University Press.

Stearns S.C. 1976. Life-history tactics: a review of the ideas. *Quart. Rev. Biol.* **51**:3–47.

Stearns S.C. 1977. The evolution of life history traits: a critique of the theory and a review of the data. *Ann. Rev. Ecol. Syst.* **8**:145–171.

Stebbins G.L. 1950. *Variation and Evolution in Plants*. New York, Columbia University Press.

Stebbins G.L. 1957. Self-fertilization and population variability in the higher plants. *Amer. Nat.* **91**:337–354.

Stebbins G.L. 1958. Longevity, habitat, and release of genetic variability in the higher plants. *Cold Spring Harbor Symp. Quant. Biol.* **23**:365–378.

Stebbins G.L. 1970. Variation and evolution in plants: progress during the past twenty years. In Hecht M.K. and W.C.Steere (eds.), *Essays in Evolution and Genetics,* pp. 173–208. Amsterdam, North-Holland.

Stebbins G.L. 1974. *Flowering Plants*. Cambridge, Harvard University Press.

Steenis C.G.G.J.van, 1958. Rejuvenation as a factor for judging the status of vegetation types. The biological nomad theory. In *Proceedings of the Symposium on Humid Tropics Vegetation, Kandy*. Paris, UNESCO.

Stern W.R. 1965. The effect of density on the performance of individual plants in subterranean clover swards. *Aust. J. Agric. Res.* **16**:541–555.

Steward K.K. and W.H.Ornes, 1975. The

autecology of sawgrass in the Florida everglades. *Ecology* **56:**162–171.

Stinson H.T. and D.N.Moss, 1960. Some effects of shade upon corn hybrids tolerant and intolerant of dense planting. *Agron. J.* **52:**482–484.

Straw R.M. 1972. A Markov model for pollinator constancy and competition. *Amer. Nat.* **106:**597–620.

Struik G.J. 1965. Growth patterns of some native annual and perennial herbs in southern Wisconsin. *Ecology* **46:**401–420.

Struik G.J. and J.T.Curtis, 1962. Herb distribution in an *Acer saccharum* forest. *Amer. Midl. Nat.* **68:**285–296.

Suneson C.A. 1949. Survival of four barley varieties in mixture. *Agron. J.* **41:**459–461.

Syme J.R. and P.M.Bremmer, 1968. Growth and yield of pure and mixed crops of oats and barley. *J. Appl. Ecol.* **5:**659–674.

Synnott T.J. 1975. *Factors affecting the regeneration and growth of seedlings of* Entandrophragma utile (Dawe & Sprague) Sprague. Ph.D. Thesis, Makerere University, Kampala.

Swart E.R. 1963. Age of the Baobab tree. *Nature* **198:**708–709.

Tadaki Y. 1963. The pre-estimating of stem yield based on the competition-density effect. *Gov. For. Expt. Stn. Tokyo Bull.* **154.**

Tadaki, Y. 1964. Effect of thinning on stem volume yield studied with competition-density effect. On the case of *Pinus densiflora. Gov. For. Expt. Stn. Tokyo Bull.* **166.**

Tadaki Y. and T.Shidei, 1959. Studies on the competition of forest trees. II. The thinning experiment on small model stand of Sugi *(Cryptomeria japonica)* seedlings. *J. Jap. For. Soc.* **41:**341–349.

Tamm C.O. 1972a. Survival and flowering of some perennial herbs. II. The behavior of some orchids on permanent plots. *Oikos* **23:**23–28.

Tamm C.O. 1972b. Survival and flowering of some perennial herbs. III. The behavior of *Primula veris* on permanent plots. *Oikos* **23:**159–166.

Taylorson, R.B. 1970. Changes in dormancy and viability of weed seeds in soils. *Weed Sci.* **18:**265–269.

Taylorson R.B. and S.B.Hendricks, 1977. Dormancy in seeds. *Ann. Rev. Pl. Physiol.* **28:**331–354.

Temple A. 1977. *Ericaceae:* polymorphisme architectural d'une famille dans régions tempérées et tropicales d'altitude. *C.R. Acad. Sc. Paris* 284 Sér. D., 163–166.

Tenhunen J.D., J.A.Weber, L.H.Filipek and D.M.Gates, 1977. Development of a photosynthesis model with an emphasis on ecological applications. III. Carbon dioxide and oxygen dependencies. *Oecologia* **30:**189–207.

Tenhunen J.D., C.S.Yocum and D.M.Gates, 1976. Development of a photosynthesis model with an emphasis on ecological applications. I. Theory. *Oecologia* **26:**89–100.

Tepper H.B. and G.T.Bamford, 1960. Thinning sweetgum stands in southern New Jersey. *USDA For. Serv., NE For. Expt. Stn., For. Res. Note* **95.**

Tevis L. 1958a. Germination and growth of ephemerals induced by sprinkling a sandy desert. *Ecology* **39:**681–688.

Tevis L. 1958b. Interelations between the harvester ant, *Veromessor pergandei* (Mayr) and some desert ephemerals. *Ecology* **39:**695–704.

Thomas A.G. 1974. Reproductive strategies of *Hieracium.* AIBS Meetings Plant Population Dynamics Symposium, Tempe, Arizona.

Thomas A.G. and H.M.Dale, 1974. Zonation and regulation of old pasture populations of *Hieracium floribundum. Can. J. Bot.* **52:**1451–1458.

Thomas A.G. and H.M.Dale 1975. The role of seed reproduction in the dynamics of established populations of *Hieracium floribundum* and a comparison with that of vegetative reproduction. *Can. J. Bot.* **53:**3022–3031.

Thomas A.G. and H.M.Dale 1976. Cohabitation of three *Hieracium* species in relation to the spatial heterogeneity in an old pasture. *Can. J. Bot.* **54:**2517–2529.

Thompson P.A. 1970. A comparison of the germination character of species of *Caryophyllaceae* collected in Central Germany. *J. Ecol.* **58:**699–711.

Thomson A.J. 1970. Analysis of yield from a competition trial with perennial ryegrass. *J. Brit. Grassl. Soc.* **25:**309–313.

Thomson G.M. 1880. On the fertilization, etc., of New Zealand flowering plants. *Trans. N.Z. Inst.* **13:**241–291.

Thornley J.H.M. 1977. A model of apical bifurcation applicable to trees and other organisms. *J. Theor. Biol.* **64:**165–176.

Thullen R.J. and P.E.Keeley, 1975. Yellow nutsedge sprouting and resprouting potential. *Weed. Sci.* **23:**333–337.

Thurston J.M. 1964. Weed studies in winter

wheat. *Proc. 7th Brit. Weed Control Conf.* pp. 592–598.

Tomlinson P.B. 1974. Breeding systems in trees native to tropical Florida—a morphological assessment. *J. Arnold Arb.* **55**: 269–290.

Tomlinson P.B. and F.C.Craighead, 1972. Growth ring studies on the native trees of sub-tropical Florida. In Ghouse A.K.M. and A.M.Gill (eds.), *Research Trends in Plant Anatomy*, pp. 39–51. New Delhi, Tata McGraw-Hill.

Tomlinson P.B. and A.M.Gill, 1973. Growth habits of tropical trees: some guiding principles. In Meggers B.J., E.S.Ayensu and W.D.Duckworth (eds.), *Tropical Forest Ecosystems in Africa and South America: A Comparative Review*, pp. 129–143. Washington, Smithsonian Institution Press.

Tomlinson P.B. and P.K.Soderholm, 1975. The flowering and fruiting of *Corypha elata* in South Florida. *Principes* **19**:83–90.

Toole E.H. and E.Brown, 1946. Final results of the Duvel buried seed experiment. *J. Agric. Res.* **72**:201–210.

Toumey J.W. and C.F.Korstian, 1937. *Foundations of Silviculture upon an Ecological Basis.* New York, Wiley.

Trenbath B.R. 1974. Biomass productivity of mixtures. *Adv. Agron.* **26**:177–210.

Trenbath B.R. 1976. Plant interactions in mixed crop communities. In Papendick, R.I., P.A. Sanchez and G.B.Triplet (eds.). *Multiple Cropping*, pp. 129–169. Madison, Amer. Soc. Agronomy.

Trenbath B.R. 1977. Interactions among diverse hosts and diverse parasites. *Ann. N.Y. Acad. Sci.* **287**:124–150.

Troll, W. 1937–43. *Vergleichende Morphologie der Höheren Pflanzen.* Berlin, Borntrager.

Tubbs, C.H. 1977. Age and structure of a northern hardwood selection forest, 1929–1976. *J. For.* **75**:22–24.

Uranov A.A. 1967. *Ontogenesis and Age Composition of Populations of Flowering Plants.* Moscow, Nauka. (In Russian.)

Uranov A.A. 1968. *Problems of Morphogenesis of Flowering Plants and the Structure of their Populations.* Moscow, Nauka. (In Russian.)

Uranov A.A. 1974. *The Age Composition of Populations of Flowering Plants in relation to their Ontogenesis.* Moscow, MGPI. (In Russian.)

Valdeyron G., B.Dommée and P.Vernet, 1977. Self-fertilization in male-fertile plants of a gynodioecious species: *Thymus vulgaris. Heredity* **39**:243–250.

Vandermeer J.H. 1977a. Notes of density dependence in *Welfia georgii* Wendl. ex Burret (Palmae) a lowland rain forest species in Costa Rica. *Brenesia* **10/11**:9–15.

Vandermeer, J.H. 1977b. Seed dispersal of a common Costa Rican rain forest palm *(Welfia georgii). Ecology* (in press).

Vandermeer, J.H. 1977c. Hoarding behavior of captive *Heteromys desmarestianus*, on the fruits of *Welfia georgii*, a rain forest dominant from Costa Rica. *Journal of Mammalogy* (in press).

Vandermeer J.H., J.Stout and G.Miller, 1974. Growth rates of *Welfia georgii, Socratea durissima* and *Iriartea gigantea* under various conditions in a natural rain forest in Costa Rica. *Principes* **18**:148–154.

van der Pijl L. 1969. *The Principles of Dispersal in Higher Plants.* Berlin, Springer-Verlag.

van der Pijl L. 1972. *Principles of Dispersal in Higher Plants.* 2nd edition. New York, Springer-Verlag.

van der Pijl L. and C.H.Dodson, 1966. *Orchid Flowers: their Pollination and Evolution.* Coral Gables, University of Miami Press.

van der Plank J.E. 1963. *Plant Diseases: Epidemics and Control.* New York, Academic Press.

van der Valk A.G. and C.B.Davis, 1976. The seed banks of prairie glacial marshes. *Can. J. Bot.* **54**:1832–1838.

Van Valen L. 1975. Life, death and energy of a tree. *Biotropica* **7**:260–269.

Vasek F.C. 1968. The relationships of two ecologically marginal sympatric *Clarkia* populations. *Amer. Nat.* **102**:25–40.

Vasek F.C. and J.Harding, 1976. Outcrossing in natural populations. V. *Evolution* **30**:403–411.

Villiers, T.A. 1972. Cytological studies in dormancy. II. Pathological ageing changes during prolonged dormancy and recovery upon dormancy release. *New Phytol.* **71**:145–152.

Villiers T.A. 1973. Ageing and the longevity of seeds in field conditions. In Heydecker W. (ed.), *Seed Ecology*, pp. 265–388. University Park, Pennsylvania State University Press.

Villiers T.A. 1974. Seed ageing: chromosome stability and extended viability of seeds stored fully imbibed. *Plant Physiol.* **53**:875–878.

Vizcaíno M. and J.Sarukhán (in preparation) Temporal and spatial patterns of leaf litter by species in a tropical deciduous forest in the Pacific Coast of Mexico.

Vogl R.J. 1967. Fire adaptations of some southern California plants. *Proc. 7th Ann. Tall Timbers Fire Ecol. Conf.* 79–109.

Vogl R.J. 1969. The role of fire in the evolution of the Hawaiian flora and vegetation. *Proc. 9th Ann. Tall Timbers Fire Ecol. Conf.* 5–60.

Vogl R.J. 1974. Effects of fire on grasslands. In Kozlowski T.T. and C.E.Ahlgren (eds.), *Fire and Ecosystems,* pp. 139–194. New York, Academic Press.

Wallace A.R. 1876. *The Geographical Distribution of Animals.* London, Macmillan.

Wallace, B. 1955. Inter-population hybrids in *Drosophila melanogaster. Evolution* 9:302–316.

Wallace B. 1963. Modes of reproduction and their genetic consequences. In Hanson W.D. and H.F.Robinson (eds.), *Statistical Genetics and Plant Breeding,* pp. 3–17. Washington D.C., Nat. Acad. Sci. (NRC, publ. 982).

Wareing P.F. 1966. The ecological aspects of seed dormancy and germination. In Hawkes J. (ed.), *Reproductive Biology and Taxonomy of Vascular Plants,* pp. 103–121. New York, Pergamon.

Watt A.S. 1947. Contributions to the ecology of bracken. IV. The structure of the community. *New Phytol.* 46:97–121.

Weaver J.E. and F.E.Clements, 1938. *Plant Ecology,* 2nd edition. New York, McGraw-Hill.

Webb L.J. 1958. Cyclones as an ecological factor in tropical lowland rain forest, north Queensland. *Aust. J. Bot.* 6:220–228.

Webb L.J., J.G.Tracey, W.T.Williams and G.N.Lance, 1967. Studies in the numerical analysis of complex rain-forest communities. II. The problem of species sampling. *J. Ecol.* 55:525–538.

Webb L.J., J.G.Tracey and W.T.Williams, 1972. Regeneration and pattern in the subtropical rain forest. *J. Ecol.* 60:675–695.

Webb R.A. 1971. The relationship of seed number to berry weight in black currant fruit. *J. Hort. Sci.* 46:147–152.

Webb R.A. 1972. Use of the boundary line in the analysis of biological data. *J. Hort. Sci.* 47:309–319.

Went F.W. 1948. Ecology of desert plants. I. Observations on germination in the Joshua Tree National Monument, California. *Ecology* 29:242–253.

Went F.W. 1949. Ecology of desert plants. II. The effects of rain and temperature on germination and growth. *Ecology* 30:1–13.

Went F.W. 1957. *The Experimental Control of Plant Growth.* Waltham, Mass., Chronica Botanica Co.

Went F.W. 1971. Parallel evolution. *Taxon* 20:197–226.

Went F.W. 1973. Competition among plants. *Proc. Nat. Acad. Sci. USA* 70:585–590.

Went F.W. and M.Westergaard, 1949. Ecology of desert plants. III. Development of plants in the Death Valley National Monument, California. *Ecology* 30:26–38.

Werner P.A. 1975a. A seed trap for determining patterns of seed deposition in terrestrial plants. *Can. J. Bot.* 53:810–813.

Werner P.A. 1975b. Predictions of fate from rosette size in Teasel (*Dipsacus fullonum* L.). *Oecologia* 20:197–201.

Werner P. 1977. Ecology of plant populations in successional environments. *Syst. Bot.* 1:246–268.

Werner, P.A. 1979. Niche differentiation and coexistence in plants: the role of seeds and seedlings. In Solbrig O.T., S.Jain, G.Johnson and P.H.Raven (eds.), *"Topics in Plant Population Biology"*, New York, Columbia University Press.

Werner P.A. (in prep.). Asexual reproduction in higher plants. *Ann. Rev. Ecol. Syst.*

Werner P.A. and W.J.Platt, 1976. Ecological relationships of co-occurring goldenrods (*Solidago:* Compositae). *Amer. Nat.* 110:959–971.

Wesson G. and P.F.Wareing, 1969a. The role of light in the germination of naturally occurring populations of buried weed seeds. *J. Exp. Bot.* 20:402–413.

Wesson G. and P.F.Wareing, 1969b. The induction of light sensitivity in weed seeds by burial. *J. Exp. Bot.* 20:413–425.

Westoby M. 1976. Self-thinning in *Trifolium subterraneum* not affected by cultivar shape. *Aust. J. Ecol.* 1:245–247.

Westoby M. 1977. Self-thinning driven by leaf area not by weight. *Nature* 265:330–331.

Whigham D. 1974. An ecological life history study of *Uvularia perfoliata* L. *Amer. Midl. Nat.* 91:343–359.

White J. 1977. Generalization of self-thinning of plant populations. *Nature* 268:373.

White J. 1979. The plant as a metapopulation. *Ann. Rev. Ecol. Syst.* 10 (in press).

White J. and J.L.Harper, 1970. Correlated changes in plant size and number in plant populations. *J. Ecol.* 58:467–485.

Whitehead D.R. 1969. Wind pollination in the angiosperms: evolutionary and environmental considerations. *Evolution* 23:28–35.

Whitehouse, H.L.K. 1950. Multiple-allelomorph incompatibility of pollen and style in the evolution of the angiosperms. *Ann. Bot.* **14**:199–216.

Whitford P.B. 1951. Estimation of the ages of forest stands in the prairie-forest border region. *Ecology* **32**:143–147.

Whitmore T.C. 1974. *Change with time and the role of cyclones in tropical rain forest on Kolombangara, Solomon Islands*. Commonwealth Forestry Institute. Paper 46.

Whitmore T.C. 1975. *Tropical Rain Forests of the Far East*. Oxford, Oxford University Press.

Whitney G.G. 1976. The bifurcation ratio as an indicator of adaptive strategy in woody plant species. *Bull. Torrey Bot. Club* **103**:67–72.

Whittaker R.H. 1975. *Communities and Ecosystems*, 2nd edition. New York, Macmillan.

Wilbur H.M. 1976. Life history evolution in seven milkweeds of the genus *Asclepias*. *J. Ecol.* **64**:223–240.

Wilbur H.M., D.W.Tinkle and J.P.Collins, 1974. Environmental uncertainty, trophic level, and resource availability in life history evolution. *Amer. Nat.* **108**:805–807.

Willey R.W. and S.B.Heath, 1969. The quantitative relationships between plant population and crop yield. *Adv. Agron.* **21**:281–321.

Willey R.W. and D.S.O.Osiru, 1972. Studies on mixtures of maize and beans (*Phaseolus vulgaris*) with particular reference to plant population. *J. agr. Sci., Camb.* **79**:517–529.

Williams G.C. 1975. *Sex and Evolution*. Princeton, Princeton University Press.

Williams J.T. and J.L.Harper, 1965. Seed polymorphism and germination. I. The influence of nitrates and low temperatures on the germination of *Chenopodium album*. *Weed Res.* **5**:141–150.

Williams O.B. 1970. Population dynamics of two perennial grasses in Australian semi-arid grassland. *J. Ecol.* **58**:869–875.

Williams W. 1951. Genetics of incompatibility in alsike clover, *Trifolium hybridum*. *Heredity* **5**:51–73.

Williams W. and A.G.Brown, 1956. Genetic response to selection in cultivated plants: gene frequencies in *Prunus avium*. *Heredity* **10**:237–245.

Williams W.T., G.N.Lance, L.J.Webb, J.G.Tracey and J.H.Connell, 1969. Studies in the numerical analysis of complex rain-forest communities. IV. A method for the elucidation of small scale forest pattern. *J. Ecol.* **57**:635–654.

Williamson A.W. 1913. Cottonwood in the Mississippi Valley. *USDA Bull.* **24**.

Wilson B.F. 1966. Development of the shoot system of *Acer rubrum* L. *Harvard Forest Paper* **14**.

Wilson B.J. 1972. Studies of the fate of *Avena fatua* seeds in cereal stubble, as influenced by autumn treatment. *Proc. 11th Br. Weed Control Conf.* 242–247.

Wilson B.J. and G.W.Cussans, 1975. A study of the population dynamics of *Avena fatua* as influenced by straw burning seed shedding and cultivation. *Weed Res.* **15**:249–258.

Winkler H. 1908. Über parthenogenesis und apogamie im Pflanzenreiche. *Progr. Rei Bot.* **2**:293–454.

Wong F.O. 1973. A study of the growth on the main commercial species in the Segaliud-Lokan F.R. Sandakan, Sabah. *Malay. Forester* **36**:20–31.

Wood B.W., S.B.Carpenter and R.F.Wittwer, 1976. Intensive culture of American sycamore in the Ohio River valley. *For. Sci.* **22**:338–342.

Wood G.H.S. 1956. The dipterocarp flowering season in North Borneo, 1955. *Malay. Forester* **19**:193–201.

Workman P.L. and S.K.Jain, 1966. Zygotic selection under mixed random mating and self-fertilization: theory and problems of estimation. *Genetics* **54**:159–171.

Wright S. 1943. Isolation by distance. *Genetics* **28**:114–138.

Wright S. 1965. The interpretation of population structure by F-statistics with special regard to systems of mating. *Evolution* **19**:395–420.

Wyatt-Smith, J. 1955. Changes in composition in natural plant succession. *Malay. Forester* **21**:185–193.

Wynne-Edwards V.C. 1962. *Animal Dispersion in Relation to Social Behaviour*. Edinburgh, Oliver and Boyd.

Yoda K., T.Kira, H.Ogawa and K.Hozumi, 1963. Intraspecific competition among higher plants. XI. Self-thinning in overcrowded pure stands under cultivated and natural conditions. *J. Biol. Osaka City Univ.* **14**:107–129.

Taxon Index

Subject Index